Enabling Methodologies for Renewable and Sustainable Energy

This book aims to provide practical aspects of, and an introduction to, the applications of various technological advancement tools, such as AI, machine learning to design, big data, cloud computing, and IoT, to model, characterize, optimize, forecast, and do performance prediction of renewable energy exploitation. It further discusses new avenues for energy sources such as hydrogen energy generation and energy storage technologies including existing policies and case studies for a better understanding of renewable energy generation.

Features:

- Covers technologies considered to explore, predict, and perform operation and maintenance of renewable energy sources.
- Aids in the design and use of renewable energy sources, including the application of artificial intelligence in a real-time environment.
- Includes IoT, cloud computing, big data, smart grid, and different optimization techniques for resource forecasting, installation, operation, and optimization of energy.
- Discusses the principle of integration/hybridization of renewable energy sources along with their optimization based on energy requirements.
- Reviews the concepts and challenges involved in the implementation of smart grids.

This book is aimed at researchers and graduate students in renewable energy engineering, computer and mechanical engineering, novel technologies, and intelligent systems.

Enabling Methodologies for Renewable and Sustainable Energy

Edited by

Gaurav Saini, Ramani Kannan,
Ernesto Benini and Krishna Kumar

CRC Press
Taylor & Francis Group
Boca Raton London New York

CRC Press is an imprint of the
Taylor & Francis Group, an **informa** business

Designed cover image: © Shutterstock

First edition published 2023
by CRC Press
6000 Broken Sound Parkway NW, Suite 300, Boca Raton, FL 33487-2742

and by CRC Press
4 Park Square, Milton Park, Abingdon, Oxon, OX14 4RN

CRC Press is an imprint of Taylor & Francis Group, LLC

ISBN: 9781032224763 (hbk)
ISBN: 9781032224787 (pbk)
ISBN: 9781003272717 (ebk)

DOI: 10.1201/9781003272717

Typeset in Times
by codeMantra

Enabling Methodologies for Renewable and Sustainable Energy

Edited by

Gaurav Saini, Ramani Kannan,
Ernesto Benini and Krishna Kumar

CRC Press
Taylor & Francis Group
Boca Raton London New York

CRC Press is an imprint of the
Taylor & Francis Group, an **informa** business

Designed cover image: © Shutterstock

First edition published 2023
by CRC Press
6000 Broken Sound Parkway NW, Suite 300, Boca Raton, FL 33487-2742

and by CRC Press
4 Park Square, Milton Park, Abingdon, Oxon, OX14 4RN

CRC Press is an imprint of Taylor & Francis Group, LLC

ISBN: 9781032224763 (hbk)
ISBN: 9781032224787 (pbk)
ISBN: 9781003272717 (ebk)

DOI: 10.1201/9781003272717

Typeset in Times
by codeMantra

Contents

Editors

Gaurav Saini is presently working as the Professor (Assistant) in the Department of Mechanical Engineering, Harcourt Butler Technical University Kanpur, Uttar Pradesh. He has Post-Doctoral Fellow experience with the Department of Sustainable Energy Engineering, Indian Institute of Technology Kanpur. Prior to joining IIT Kanpur, he was serving as the Professor (Assistant) in the School of Advanced Materials, Green Energy and Sensor Systems, Indian Institute of Engineering Science and Technology (IIEST) Shibpur, India. He received his Ph.D. in Turbomachines (Hydrokinetic Turbines) in the year 2020 and M. Tech (Fluid Machinery and Energy Systems) in the year 2014 from the Indian Institute of Technology Roorkee, Uttarakhand, India.

After his Ph.D. from IIT Roorkee, he was working as project fellow in the department of hydro and renewable energy, Indian Institute of Technology Roorkee. His research areas include renewable energy (hydrokinetic energy, wind power, and biomass), Computational Fluid Dynamics (CFD) and fluid mechanics and turbomachines; fluid power. He has published several research publications on renewable energy technologies in different international journals of repute. He has also presented his research at different international and national platforms and he received accolades from various peers working in the same area across the globe. He is skilled in computational fluid dynamics (CFD) – numerical modelling and roto-dynamics analysis, multiphase flow analysis, modeling of various renewable energy resources viz. wind, marine, solar and hydrokinetic energy for rural applications, wind and hydrokinetic – technology selection and design, installation strategies, performance evaluation and O&M issues.

Ramani Kannan is a Senior Lecturer in Universiti Teknologi PETRONAS, Malaysia. He received his BEng degree from Bharathiar University, India. Later, he completed his MEng and PhD in Power Electronics and Drives in Anna University. He holds more than 125 publications in reputed international and national journals and conferences; also, he has published six books. He has completed five funded projects and has seven research projects in progress. The grants are FRGS, ASEAN-India, YUTP, KETTHA, and STIRF. He is a chartered engineer (CEng, the UK), active senior member of IEEE (the USA), and member of IE (I), IET (the UK), ISTE (I), and Institute of Advanced Engineering and Science (IAENS). Dr. Ramani is recognized with many awards, including "Career Award for Young Teacher" from AICTE India, 2012; "Young Scientist Award" in power electronics and drives, 2015; "Highest Research Publication Award" 2017; Award for Outstanding Performance, Service and Dedication 2019 at UTP, Malaysia; "Outstanding Researcher Award" at UTP, Q Day 2019; and Best Presenter Award, IEEE CENCON 2019 international conference at Indonesia. He actively serves as Secretary, IEEE Power Electronics Society, Malaysia, since 2020. He is the Editor-in-Chief of the *Journal of Asian Scientific Research* (2011–2018) and Regional Editor of the *International Journal of Computer-Aided Engineering and Technology*, Inderscience Publishers, the UK, from 2015. He is an Associate Editor of the IEEE Access journal since 2018. He is Editor of *International Transactions on Electrical Energy Systems* (ITEES Wiley) since 2021 and is Section Editor of *Platform: A Journal of Engineering*, UTP Press, MY, since 2020. He is an Associate Editor of the *Advanced Materials Science and Technology* journal since 2020. Dr. Ramani serves as a guest editor for many publishers such as Elsevier, Inderscience, IGI Global, CRC, Taylor & Francis, Bentham Science, and IJPAM. His research interest involves power electronics, inverters, modeling of induction motor, AI, machine learning, and optimization techniques.

Ernesto Benini graduated with honors in Mechanical Engineering (1996) and obtained his PhD in Energy Technology (2000). He is currently a Professor of "Fluid Machinery" at the Department of Industrial Engineering of the University of Padua, Italy. He has more than 20 years of experience in the research and development of advanced methods for the design and optimization of fluid machines involving renewables. He is responsible for numerous private and European research projects and is a scientific advisor for major international companies in the renewable energy sector. He is a senior member of ASME and AIAA. He serves as Editor of authoritative scientific journals such as ASME *Journal of Engineering for Gas Turbines and Power* and *International Journal of Turbo & Jet Engines*. He has authored and co-authored about 300 scientific publications concerning the fluid dynamics of machines. He has received numerous awards for his academic and scientific activities, including numerous "Best Paper Awards" and the gold medal for the Best Academic Curriculum, and was nominated for the New York Academy of Sciences' "Outstanding Young Researchers" award.

Krishna Kumar is presently working as a Research and Development Engineer at UJVN Ltd. Before joining UJVNL, he worked as an Assistant Professor at BTKIT, Dwarahat (Uttarakhand), India. He received his B.E. (Electronics and Communication Engineering) from Govind Ballabh Pant Engineering College, Pauri Garhwal, Uttarakhand, India, and M. Tech (Digital Systems) from Motilal Nehru National Institute of Technology, Allahabad, India. He is also pursuing his PhD in the Indian Institute of Technology Roorkee, India. He has more than 12 years of industrial and teaching experience and has published numerous research papers in international journals such as IEEE, Elsevier, Taylor & Francis, Springer, and Wiley. His research area includes renewable energy, artificial intelligence, cloud computing, and the Internet of things.

Contributors

Amit Kumar Bairwa
Department of Computer Science and Engineering
Manipal University
Jaipur, India

N.P.G. Bhavani
Department of ECE
Saveetha School of Engineering, Saveetha Institute of Medical and Technical Sciences
Chennai, India

Ranchan Chauhan
Mechanical Engineering Department
Dr B R Ambedkar NIT
Jalandhar, India

Sandeep Chaurasia
Department of Computer Science and Engineering
Manipal University
Jaipur, India

Suman Dutta
Department of Electrical Engineering
Dr. K. N. Modi University
Newai, India

Adebayo B. Fakeye
Covenant University
Ota, Nigeria

Mohit Garg
Electrical Engineering Department
DCR University of Science & Technology
Sonipat, India

Manoj Gupta
JECRC University
Jaipur, India

Rajesh Gupta
Departments of Energy Science and Engineering
Indian Institute of Technology Bombay
Mumbai, India

U. Jayalatsumi
Department of ECE
Dr. M.G.R. Educational and Research Institute
Chennai, India

Sandeep Joshi
Department of Computer Science and Engineering
Manipal University
Jaipur, India

T. Kavitha
Department of Civil Engineering
Dr. MGR Educational & Research Institute
Chennai, India

Yugal Kishor
Department of Electrical Engineering
National Institute of Technology
Raipur, India

K. Krishnakumar
School of Electrical Engineering
Vels institute of Science, Technology & Advanced Studies
Chennai, India

Anil Kumar
Electrical Engineering Department
DCR University of Science & Technology
Sonipat, India

Krishna Kumar
Hydro and Renewable Energy
Indian Institute of Technology Roorkee
Uttarakhand, India

K. Senthil Kumar
Department of ECE
Dr. M.G.R. Educational and Research Institute
Chennai, India

Manish Kumar
Departments of Energy Science and Engineering
Indian Institute of Technology Bombay
Mumbai, India

Narendra Kumar
School of Computing
DIT University
Dehradun, India

Naresh Kumar
Electrical Engineering Department
DCR University of Science & Technology
Sonipat, India

Satpal Singh Kushwaha
Department of Computer Science and Engineering
Manipal University
Jaipur, India

Roopmati Meena
Departments of Energy Science and Engineering
Indian Institute of Technology Bombay
Mumbai, India

D. Misra
Techno India Group
Kolkata, India

Sunday O. Oyedepo
Covenant University
Ota, Nigeria

R.N. Patel
Department of Electrical Engineering
National Institute of Technology
Raipur, India

Mohana Rajendran
Department of Civil Engineering
Mepco Schlenk Engineering College
Sivakasi, India

Roshan Raman
Department of Mechanical Engineering
The NorthCap University
Gurugram, India
Centre for Advanced Studies and Research in Automotive Engineering
Delhi Technological University
New Delhi, India

C.H. Kamesh Rao
Department of Electrical Engineering
National Institute of Technology
Raipur, India

Gaurav Saini
Harcourt Butler Technical University Kanpur
Uttar Pradesh, India

Rachna Shah
National Informatics Centre
Dehradun, India

Ashutosh Sharma
Mechanical Engineering Department
Dr B R Ambedkar NIT
Jalandhar, India

Vivek Shukla
Hydrogen Energy Center, Department of Physics, Institute of Science
Banaras Hindu University
Varanasi, India

Alok Kumar Singh
Department of Electrical Engineering
Dr. K. N. Modi University
Newai, India

Arambakam Sreeram
Department of Electrical Engineering
National Institute of Technology
Raipur, India

K. Sujatha
Department of Electrical and Electronics Engineering
Dr. M.G.R. Educational and Research Institute
Chennai, India

Neeraj Tiwari
Department of Electrical Engineering
Poornima University
Jaipur, India

Thakur Prasad Yadav
Hydrogen Energy Center, Department of Physics, Institute of Science
Banaras Hindu University
Varanasi, India

Preface

Economic development and environmental preservation are the key factors to compete and sustain in the current challenging world. The use of renewable energy sources plays a major role to meet the aforementioned objectives. Nowadays, people are more focused to utilize the vast potential of different sources of renewable energy viz. hydropower, solar, geothermal, biomass and wind. The development in renewable energy (RE) generation has sparked a paradigm change in the energy sector. In the current scenario, approximately 17% of the world energy requirement is obtained by different sources of renewable energy and this contribution is still going to increase with special emphasis to reduce the negative impacts of fossil fuel-based conventional energy sources.

However, the fluctuation in energy generation due to variability in weather conditions and availability of renewable energy resources are the key challenges to overcome. Therefore, continuous advancements in technologies and process control are being made to mitigate the variable output nature of energy generated from renewable energy sources. On the other hand, variations in the energy demand accelerate the focus towards innovative technologies to integrate with the renewable energy-based sources.

Various state-of-art tools such as artificial intelligence (AI), machine learning algorithms, analytical modeling, big data, cloud computing, and internet of things (IoT) solutions may help to effectively manage and use the resources, control energy flows, regulate the grid, and optimize the work process. The use of real-time predictive analytics and data science solutions requires significant investment and readiness to face various challenges. By keeping in view, the current scenario, the use of artificial intelligence is necessary to revolutionize the energy market and to effectively harness the vast potential of renewable energy.

The present book aims to provide practical aspects and introduction to the application of various technological advancement tools such as artificial intelligence, machine learning, big data, cloud computing, and internet of things during resource forecasting, modelling of system, characterization and optimization of parameters and performance prediction of renewable energy sources.

1 Evolution of Sustainable Energy from Power Concrete Construction

Mohana Rajendran
Mepco Schlenk Engineering College

CONTENTS

1.1 OUTLINE OF SUSTAINABLE ENERGY

Major hazardous environmental issues such as global warming, greenhouse gas emission, air pollution, and depletion of natural resources raised from the burning of fossil fuel while generating electricity demand human society to switch over to renewable and sustainable energy sources. The terms renewable and sustainable energy are most commonly used worldwide as the initiation of the environment-friendly mode of energy generation and consumption. Many researchers have used both terms interchangeably, but actually, they are not. The term renewable means restorable energy even after its consumption (Wang et al., 2018). In contrast, the term sustainable energy fulfills the demands without depletion and any hazardous effects to the environment. Also, to achieve long-term sustainability, economic, social, and environmental parameters should be equally satisfied. Not all renewable energy resources need to be required to be sustainable and vice versa (Dehghanpour et al., 2019). For instance, solar and wind energy sources are renewable as they always replenish the environment. But the sustainability could not be achieved from the economic and environmental point of view due to the high installation cost and disposables after its life cycle (Yoro et al., 2021; Proto et al., 2018).

DOI: 10.1201/9781003272717-1

Similarly, nuclear energy itself is renewable, whereas the core material uranium and its reactive chain mechanism used in the power plant make nuclear energy a non-renewable, costlier, and environmentally hazardous type of resource. Also, the disposal of residual graphite tailings from nuclear power plants is a challenge nowadays (Shishegaran et al., 2020). Hence, it should be noticed that all the renewable energy sources are not necessarily sustainable. The energy sources that satisfy both renewable and sustainable considerations are the present need of research in today's world (Brodny and Tutak, 2021).

The ore waste from the graphite mining industry known as graphite tailings is used as alternative siliceous materials in autoclaved and aerated concrete to reduce the cost of pure graphite in concrete. It is observed that up to 60% of graphite tailings could be successfully used as a binder (Peng et al., 2021). Complete switching of renewable energy resources such as solar power and wind power requires a high area for installation and high initial cost investments. The major consumers of the generated electricity are the industrial sector. The automobile industry consumes nearly 27%–35% of the total generated electricity per year for operation and maintenance purposes. The annual disposal rate of automobile wastes exceeds its consumption rate (Kontoleon et al., 2013; Zhao et al., 2019). Recent studies carried out on the utilization of residual waste from the iron ores suggested that 40% replacement of iron ore wastes as fine aggregate resulted in superior strength properties in the ultra-high-performance concrete. Also, the addition of iron ore residual wastes ranging from 40% to 80% by weight of fine aggregate enhanced the impermeability characteristics and led to the formation of highly durable and serviceable energy-efficient concrete structures (Zhang et al., 2020; Ullah et al., 2021).

The current ongoing project in Lancaster University, London, is focused on the development of potassium-based geopolymer concrete as a conductive medium to store and deliver electricity at room temperature and as a self-sensing concrete to assess the strain and stress developed due to the applied stress without any admixtures and electronic sensor. Both functions of electricity storage and self-sensing mechanism have been achieved only by means of diffusion in the potassium ions. This potassium-based geopolymer system is planned to be implemented as a power concrete structure that can store energy like battery and smart self-sensing concrete member in case of bridge structures, building elements, and pavements after completing research works (Saafi et al., 2018; Li et al., 2019). Similarly, the residual iron ore waste from the steel industry is employed as an alternate fine aggregate system, which resulted in the enhancement of strength and thermal conductivity properties (Nóbrega Mendes et al., 2020; Karimi et al., 2021). The use of residual iron ore wastes in the construction industry has been proved to be a cost-effective mode of conductive medium that can be suggested to the applications of chemical factories and de-icing agents in the colder regions (Liu et al., 2015; Fulham-Lebrasseur et al., 2020). The major engine scrap waste components consist of spark plugs and cast-iron engine blocks, which have an excellent heat conducting behavior. The cast-iron sheets from engine blocks can withstand up to 165°C and conduct a large amount of heat. The spark plug functions to prevent the bursting of the engine by allowing controlled explosion. The spark plug is a composite material, which consists of iridium and ceramic layers. The iridium acts as an anti-corrosion filament, and the ceramic layer acts as an insulating material (Chess and Green, 2019; Charilaou et al., 2021).

The research work carried out in the central station of Sweden proved that the body temperature generated from the passengers could be sufficient to warm the building and to generate electricity in a sustainable way. The method is now successful in warming the neighboring building. On average, the heat generated from 250,000 travelers per day is used for electricity generation (Lucio-Martin et al., 2021; Berardi Andres and Gallardo, 2019). This chapter mainly aims to investigate the efficiency of this sustainable energy generation practically and establish an economical and efficient alternative way of electricity generation from scrap waste materials through infrastructural development projects and curb the use of non-renewable energy resources.

1.2 FABRICATION METHOD

The theme is extracted from the concept of the greenhouse gas effect. It is a well-known phenomenon that greenhouse gases in the atmosphere trap the energy from sunlight and make the earth warmer. This concept is implemented internally in the proposed composite concrete system. The concrete consists of three layers: top spark plug layer, middle concrete, and bottom cast iron layer. All the conductive materials are derived from the waste engine scrap materials. The internal greenhouse effect is achieved by redirecting the absorbed heat energy between the spark plug and cast iron sheet layer by means of thermal conductivity of the middle concrete layer as shown in Figure 1.1. To enhance the thermal conductivity of the concrete, iron ore slag wastes are used as a partial replacement material for fine aggregate. The conducted heat energy is trapped by cast iron sheets and extracted using embedded copper wires. The copper wires are soldered within the waste high-density polyethylene (HDPE) duct pipe installed between the bottom cast ion sheets. The wires are connected to the thermoelectric generator, and the efficiency of the generated electricity has been assessed by connecting to the light-emitting diode (LED) setup.

1.2.1 Composite Layer

Cast iron sheets from the waste engine body are cut in the size of 2×2 inches and laid at the sides of the pipe. This cast iron layer is used to provide efficient storing and easy heat conduction to the copper wires. The installation of copper wires is done by means of soldering into the HDPE duct pipe as shown in Figure 1.2.

For long-term energy-storing purposes, 40% of fine aggregate has been replaced by iron ore slag, which possesses good strength and conductivity properties. In the top one-third, finely grounded spark plugs are filled to trap sufficient heat energy and resist the surface's heat radiation. The spark plug consists of iridium and ceramic portions. The ceramic portions are separated and placed on the top layer in order to store a higher rate of heat energy by preventing the thermal radiation effect. Iridium plays an important role in protecting the specimen from corrosion by offering an anti-rusting effect (Attar et al., 2020).

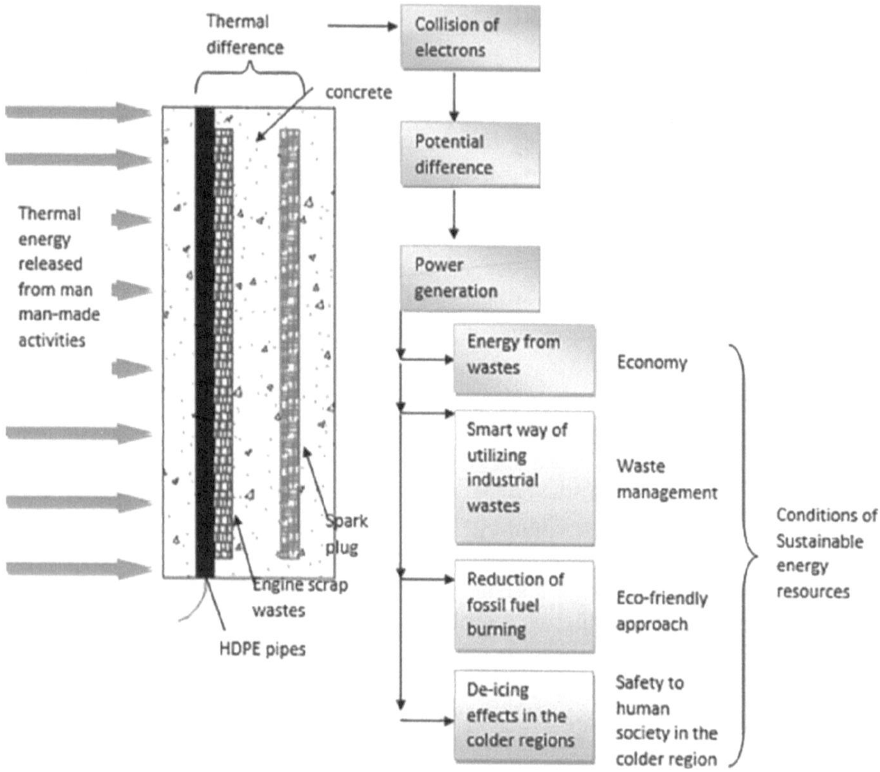

FIGURE 1.1 Prototype for sustainable power generation.

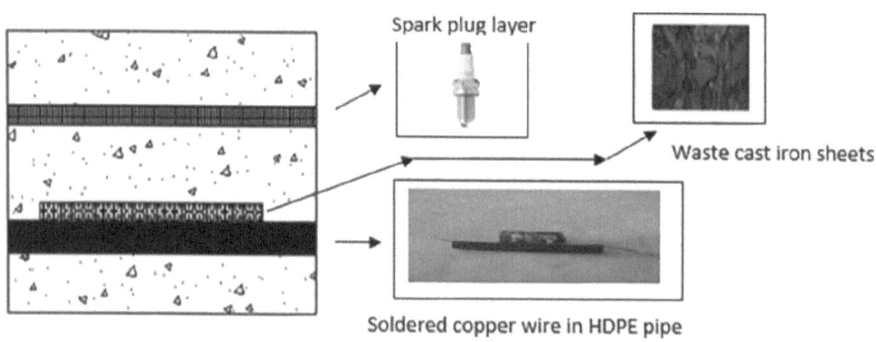

FIGURE 1.2 Details of the copper wire arrangements in the specimen (model).

1.2.2 Pre-Engineered Arrangements

The iron sheets from the engine scrap materials are cut using auger blades in $5 \times 5 \times 0.35$ cm. The iron sheets are immersed for 15 minutes in the hydrochloric acid solution diluted with distilled water in the ratio of 1:4 proportions. After the complete emission of pungent-smelling gas, the iron sheets are cooled in distilled water for 15 minutes. Thus, the rust formed on the surface is completely washed away and the pure layer of cast iron is exposed. The HDPE pipe is cut in the size of 150 mm for the entire length distribution in the cube specimen. By using hexagonal drive salt auger rod drilling, the holes are provided in the size of 3 mm on the HDPE pipe.

The spark plug wastes are broken into smaller pieces by continuous mechanical blowing. The ceramic part is used as a shielding layer for preventing heat radiation. The iridium pole is powdered and mixed with fine aggregate to enable the specimen to act as an anti-rusting element. The iron ore slag from camshaft waste was rubbed against a hard surface to reduce the size. The soldered copper wires are connected to the corner of the copper sheet, which dissipates the heat to the TEG pasted over its surface. The TEG converts thermal energy into electrical energy based on the Peltier effect principle (Zhang et al., 2021; Tian et al., 2019).

1.3 CHARACTERISTICS ASSESSMENT

1.3.1 Storage of Thermal Energy

A concrete tile of 200 × 200 × 15 mm was placed in the thermal conductivity test apparatus and clamped at the ends. The concrete tile has been covered with envelope covering arrangements and a constant load difference of 80 units is maintained in between the two knobs and the results at various time intervals are shown in Figure 1.3.

After 100°C of thermal exposure, the conductivity of the proposed scrap waste concrete specimen is observed to be 5 W/mK, which is 6.25 times higher than the conventional concrete panel. The reason behind this improved thermal conductivity is the presence of voids in the concrete surface, which enables the passage of thermal

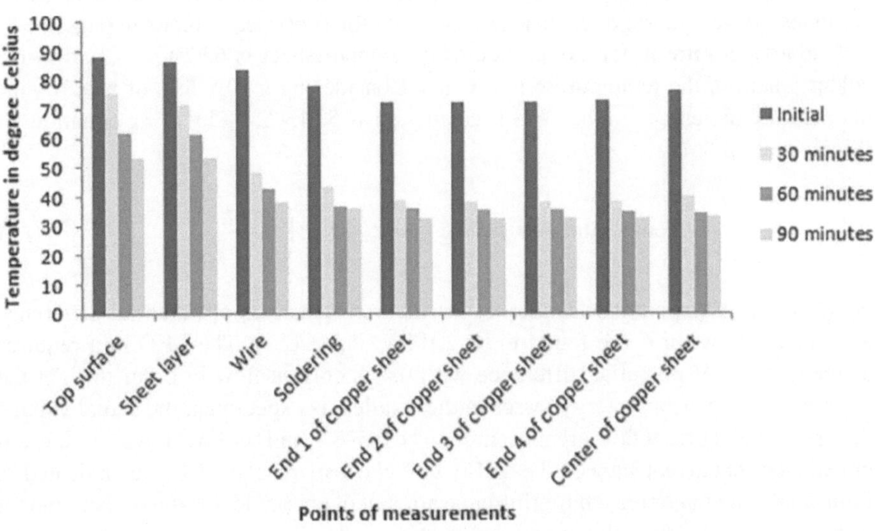

FIGURE 1.3 Comparison between the variations in thermal conductivity.

energy without any interruptions. Also, the rapid vibration of iron ore slag molecules excited by the external heat application actuated an even faster rate of heat flow (Tian and Hu, 2012). On the contrary, in the case of conventional concrete, the well-graded coarse aggregate acted as a resistance to the heat flow. The temperature storage has been varied at each connection point due to the composition of heterogeneous layers present in the proposed scrap waste concrete panels. It is observed that the maximum temperature storage is present on the top surface of the concrete. When the thermal energy is transferred to the copper sheet through the wires, the loss of thermal energy occurs due to various factors such as radiation of heat waves to the environment from the top surface, mechanical twisting of copper wires, soldering effect, and the pores present in the sides of the cube (Vijayaragavan, 2017).

1.3.2 Flow of Electricity from Sustainable Mode

The thermal storage is comparatively high in the case of iron scraps than the other parts. The thermal energy is transferred to the copper wires and the copper sheets by means of the conduction effect. As the copper wires are connected to the end of the copper sheet, the energy flow will be from corner to center. In the laboratory, to create the high thermal difference between the TEG connected to the corners of the copper sheets, two of the TEG are connected to the heat transferring copper wires and another two TEG are connected to the ice cubes as cold junction. According to the Seebeck effect, the temperature difference between the hotter and colder junctions induces the collision and fast transfer of electrons and creates the potential difference between the two terminals. This induced potential difference is converted into an electric current in the opposite direction to the electron flow. Due to this thermoelectric effect, two main things are easily achieved. One is the generation of electric current from the engine scrap wastes, and the other one is de-icing of cubes from the conducted thermal energy. The power supply induced in the connected LED setup confirms the generation of electric current from the proposed composite panels.

The temperature in the hotter side of the copper sheet is 62.7°C (t_h), and in the colder junction, the temperature is 7°C (t_c). Considering a 10% loss of heat (t_l), the temperature difference in the TEG is calculated as 50.13°C (t_d) by using equation 1.1 (Sinha et al., 2020).

$$\text{Temperature difference } t_d = (t_h - t_c) * \left(1 - \frac{t_l}{100}\right) \tag{1.1}$$

By using the interpolation technique, the theoretical values of potential difference and current flow for t_d are found to be 2.103 V and 0.42 A. The LED unit requires a supply of 1.5 V potential difference with 0.4 A current flow in order to light the system. Due to various energy losses in the smaller size specimen, the actual voltage difference and current flow are measured to be 0.578 V and 0.13 A. However, the generated electric current showed the ability to make a spark in the LED as indicated in Figure 1.4. The major reason for this huge amount of energy loss is the higher amount of heat loss points than the energy conducting nodes (He et al., 2020). This can be effectively rectified by implementing the proposed scrap waste concrete system in

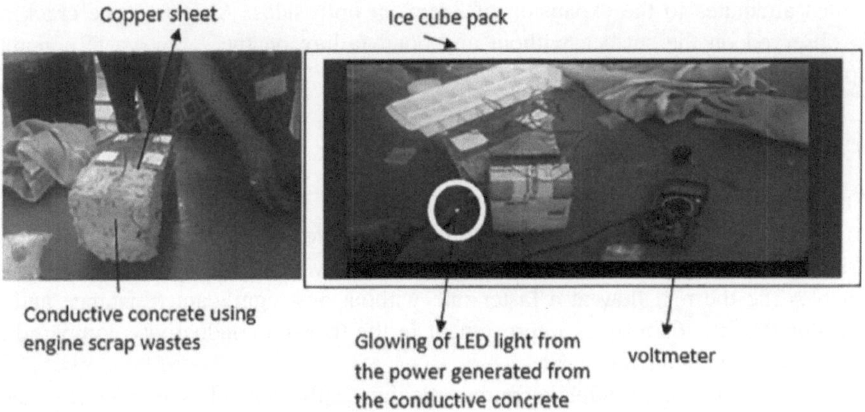

FIGURE 1.4 Assessment of the generated power from the specimen.

TABLE 1.1
Comparison of Compressive Strength

S. No.	Specimen	Maximum Load (kN)	Maximum Compressive Stress (MPa)
1.	Conventional cement concrete	616.7	27.4
2.	Scrap waste concrete	571.5	25.4

real-time mass building construction subjected to enormous amount of heat energy, especially in industrial buildings.

1.3.3 Compressive Behavior

The compressive strength test has been employed to confirm the safety measures against strength characteristics of the proposed electric power generating concrete specimen. The cube specimens of size 150 mm are tested in the compression testing machine (CTM), and the results of the scrap waste concrete are compared with the results of conventional cement mortar.

It is observed from the results of the compressive strength test that the inclusion of scrap waste layer in the concrete resulted in the reduction in compressive strength by 7.82% than the conventional concrete specimen, which is shown in Table 1.1. Air voids are provided in the concrete specimen to enhance the thermal conductivity by utilizing the gap-graded aggregates (Khamlich et al., 2021; Staley et al., 2021). This porous structure resulted in the slighter degradation of strength behavior.

The continuous application of compressive load on the specimen induced the 0.5 cm decrease in height in the conventional concrete. This leads to the striping-off phenomenon of the outer layer. Also, the fine aggregate portion is separated at the time of failure, which can be visible after the occurrence of bond breakage. However, in the case of scrap waste concrete specimen, lateral deformation is predominant, which attributes to the expansion of 3 mm on both sides. Alligator-type cracking is observed on the surface without any bond failure on the scrap waste concrete specimen.

1.4 CONCLUSIONS

Superfluous engine scrap wastes amalgamated with iron ore slag have been proved to be an efficient energy-storing medium suitable for economic and environment-friendly power concrete construction. Allowance of voids in the scrap waste concrete enables the thermal flow at a faster rate without any significant resistance and is attributed to the 6.25 times improvement in the thermal conductivity compared to the conventional concrete.

The provision of a conductive porous medium in the concrete specimen enhanced the thermal conductivity of the member without noticeable compromise in the strength characteristics. Compared to the conventional concrete, the proposed scrap waste-incorporated concrete elements showed only 7.82% degradation in the compressive strength. As the fine aggregate has been replaced by iron ore slag up to 40% by weight, the resultant sustainable concrete specimen showed alligator type of crack propagation before failure.

The conventional concrete panel identifies the bond breakage effect with external layer sheathing off and separation of fine aggregate portion at the time of failure. The storage of thermal energy increased with the increase in the depth of composite layer and was maximum at iron scrap sheets placed at the bottom of the panel. The difference in the thermal effect initiated the potential difference between the TEG and resulted in the current flow between the terminals, which is confirmed by the power supply in the LED setup.

Further development of this type of concrete using engine scrap wastes assimilated with iron ore slag composite for use in industrial building structures would be a well-organized scientific way to the sustainable mode of electricity generation for the global energy demands.

REFERENCES

Andres, U.B. and A. Gallardo. Properties of concretes enhanced with phase change materials for building applications. *Energy and Buildings*, 2019, 199, 402–414.

Attar, A., H. Lee and G.J. Snyder. Optimum load resistance for a thermoelectric generator system. *Energy Conservation and Management*, 2020, 226, 113490.

Brodny J. and M. Tutak. The comparative assessment of sustainable energy security in the Visegrad countries. A 10-year perspective. *Journal of Cleaner Production*, 2021, 317, 128427.

Charilaou, K., T. Kyratsi and L.S. Louca. Design of an air-cooled thermoelectric generator system through modelling and simulations, for use in cement industries. *Science*, 2021, 33, 101321.

Chess, P. and W. Green. Solutions for new structures. *Durability of Reinforced Concrete Structures*, 2019. CRC Press.

Dehghanpour, H., K. Yilmaz and M. Ipek. Evaluation of recycled nano carbon black and waste erosion wires in electrically conductive concretes. *Construction and Building Materials*, 2019, 221, 109–121.

Fulham-Lebrasseur, R., L. Sorelli and D. Conciatori. Development of electrically conductive concrete and mortars with hybrid conductive inclusions. *Construction and Building Materials*, 2020, 237(1–2), 117470.

He, Y., Y. Zhang, C. Zhang and H. Zhou, Energy-saving potential of 3D printed concrete building with integrated living wall. *Energy and Buildings*, 2020, 222, 110110.

Karimi, M.M., S. Amani, H. Jahanbakhsh, B. Jahangiri and A.H. Alavi. Induced heating-healing of conductive asphalt concrete as a sustainable repairing technique: A review. *Cleaner Engineering and Technology*, 2021, 4, 100188.

Khamlich, I., K. Zeng, G. Flamant, J. Baeyens, C. Zou, J. Li, X. Yang, X. He, Q. Liu, H. Yang, Q. Yang and H. Chen. Technical and economic assessment of thermal energy storage in concentrated solar power plants within a spot electricity market. *Renewable and Sustainable Energy*, 2021, 139, 110583.

Kontoleon, K.J., T.G. Theodosiou and K.G. Tsikaloudaki. The influence of concrete density and conductivity on walls' thermal inertia parameters under a variety of masonry and insulation placements. *Applied Energy*, 2013, 112, 325–337.

Li, M., T. Ma, J. Liu, H. Li, Y. Xu, W. Gu and L. Shen, Numerical and experimental investigation of precast concrete facade integrated with solar photovoltaic panels. *Applied Energy*, 2019, 253, 113509.

Liu, K., Z. Wang, C. Jin, F. Wang and X. Lu. An experimental study on thermal conductivity of iron ore sand cement mortar. *Construction and Building Materials*, 2015, 101, 932–941.

Lucio-Martin, T., M. Roig-Flores, M. Izquierdo and M.C. Alonsoa. Thermal conductivity of concrete at high temperatures for thermal energy storage applications: Experimental analysis. *Solar Energy*, 2021, 214, 430–442.

Mendes, F.N., P. Roberto, R. de Avillez, S. Letichevsky and A. Silva. The use of iron ore tailings obtained from the Germano dam in the production of a sustainable concrete. *Construction and Building Materials*, 2020, 101, 932–941.

Peng, Y., Y. Liu, B. Zhan and G. Xu. Preparation of autoclaved aerated concrete by using graphite tailings as an alternative silica source. *Construction and Building Materials*, 2021, 267, 121792.

Proto, A., D. Bibbo, M. Cerny, D. Vala, V. Kasik, L. Peter, S. Conforto, M. Schmid and M. Penhaker. Thermal energy harvesting on the bodily surfaces of arms and legs through a wearable thermo-electric generator. *Sensors*, 2018, 18, 1927.

Saafi, M., A. Gullane, B. Huang, H. Sadeghi, J. Ye, F. Sadeghi. Inherently multifunctional geopolymeric cementitious composite as electrical energy storage and self-sensing structural material. *Composite Structures*, 2018, 201(1), 766–778.

Shishegaran, A., F. Daneshpajoh, H. Taghavizade and S. Mirvaladd. Developing conductive concrete containing wire rope and steel powder wastes for route deicing. *Construction and Building Materials*, 2020, 232, 117184.

Sinha, O.P., R. Kumar and K. Dishwar, Utilization of iron ore slime and bottom ash. *Construction and Building Materials*, 2020, 378, Part A.

Staley, Z.R., C.Y. Tuan, K.M. Eskridge and X. Li. Using the heat generated from electrically conductive concrete slabs to reduce antibiotic resistance in beef cattle manure. *Science of the Total Environment*, 2021, 768, 144220.

Tian, X. and H. Hu, Test and study on electrical property of conductive concrete. *Procedia Earth and Planetary Science*, 2012, 5, 83–87.

Tian, X., F. Pei, X. Liu, L. Jia, C. Deng, X. Wang, P. Yang and H. Cheng. Feasibility study on reducing the grounding resistance of a transmission tower with conductive concrete. *Emerging Developments in the Power and Energy Industry*, 2019. CRC Press.

Ullah, S., C. Yang, L. Cao, P. Wang, Q. Chai, Y. Li, L. Wang, Z. Dong, N. Lushinga and B. Zhang. Material design and performance improvement of conductive asphalt concrete incorporating carbon fiber and iron tailings. *Construction and Building Materials*, 2021, 303, 124446.

Vijayaragavan, J. Effect of copper slag, iron slag and recycled concrete aggregate on the mechanical properties of concrete. *Resource Policy*, 2017, 53, 219–225.

Wang, C., Q. Chen, H. Fu and J. Chen. Heat conduction effect of steel bridge deck with conductive gussasphalt concrete pavement. *Construction and Building Materials*, 2018, 172, 422–432.

Yoro, K.O., M.O. Daramol, P.T. Sekoai, U.N. Wilson and O. Eterigho-Ikelegbe. Update on current approaches, challenges, and prospects of modeling and simulation in renewable and sustainable energy systems. *Renewable and Sustainable Energy Reviews*, 2021, 150, 111506.

Zhang, M., J. Wang, Y. Tian, Y. Zhou, J.Z., Huaqing, X. Zihua, W.W. Li and Y. Wang, Performance comparison of annular and flat-plate thermoelectric generators for cylindrical hot source. *Energy Reports*, 2021, 7, 413–420.

Zhang, W., X. Gu, J. Qiu, J. Liu, Y. Zhao and X. Li. Effects of iron ore tailings on the compressive strength and permeability of ultra-high-performance concrete. *Construction and Building Materials*, 2020, 260, 119917.

Zhao, R., C. Tuan, B. Luo and A. Xu. Radiant heating utilizing conductive concrete tiles. *Building and Environment*, 2019, 148, 82–95.

2 Acetylene as a Sustainable Fuel for Diesel Engine

A Case Study

Roshan Raman
The NorthCap University
Centre for Advanced Studies and Research
in Automotive Engineering

CONTENTS

2.1 INTRODUCTION

Diminution of petroleum fuels and degradation of the environment are two major challenges that the present world is facing [1]. Diesel engines are highly efficient and are used in different applications such as agriculture, transportation, and power generation [2]. Since these engines are operated majorly by conventional petroleum diesel fuel, they release harmful emissions such as NO_x (oxides of nitrogen), HCs (hydrocarbons), and particulate matter that are highly dangerous to humans, and gases such as CO_2 raise the threat of global warming [3]. Many researchers are actively working to mitigate these twin threats [4–6]. They are looking for an alternate renewable fuel that is environmentally friendly, is efficient, and can compete with diesel fuel [7,8]. Many scientists have suggested that running a conventional CI engine on dual-fuel combustion mode utilizing gaseous fuels such as LPG [9],

DOI: 10.1201/9781003272717-2

hydrogen [10–12], and natural gas [13,14] without major changes in hardware is very easy. Similar to many other gaseous fuels, acetylene can also be used in the diesel engine.

2.1.1 Acetylene Production and Properties

Conventionally, acetylene was produced from calcium carbide and the major sources of calcium carbide are limestone and coke, which are abundant in nature [15]. Nowadays, acetylene is also produced from methane by thermal cracking; thus, acetylene can be treated as a renewable fuel [16]. Acetylene is a colorless gas and has a garlic-like odor [16]. Acetylene has a high flame rate and a wide range of flammability similar to hydrogen, which can be advantageous in IC engines [2,3,17,18]. It can be easily mixed with air, thus eliminating CI engine wall wetting problems [1,19–22]. Using acetylene in diesel engines might be advantageous since it has a very high self-ignition temperature. Hence, it can be used easily even at higher compression ratios in compression ignition (CI) engines [2,5,6,8,19]. The acetylene engine avoids early combustion phasing at peak loads, which is also advantageous over homogenous charge compression ignition (HCCI) engine [2,3,5,17,20,21]. Acetylene engines approach thermodynamically ideal engine cycle at the stoichiometric condition. John Price explains the explosive nature of acetylene that may cause damage when not used properly [22]. Many scientists have highlighted the controllability of the HRR and the ignition timing as the major challenges of the HCCI engine [23,24] and have recommended utilizing the dual-fuel combustion technique to deal with the problems of HCCI mode [25–27]. Dual-fuel combustion (DFC) has been explored by many engine experts [1,2,3,5,14,28–30]. Dual-fuel engines can interchangeably use a gaseous fuel and a liquid fuel. The physicochemical properties of acetylene are compared with various fuels in Table 2.1.

TABLE 2.1
Physicochemical Characteristics of Diesel, Hydrogen, CNG, and Acetylene [1,3,5]

Properties	Diesel	Hydrogen	CNG	Acetylene
Density (kg/m^3)	840	0.08	0.754	1.092
Auto ignition temp. (K)	527	845	813	578
Stoichiometric air-fuel ratio	14.5	34.3	17.2	13.2
Flammability limits (vol%)	0.6–5.5	4.0–74	4.3–15.2	2.5–81
Flammability limits	-	0.1–6.9	0.6–1.4	0.3–9.6
Adiabatic flame temp. (K)	2200	2400	2163	2500
Lower calorific value (MJ/kg)	42.5	120	47.5	48.225
Ignition energy (mJ)	-	0.02	0.29	0.019
Flame propagation speed (m/s)	0.3	3.5	0.4	1.5

2.1.2 Dual-Fuel Engine

Rudolf Diesel is the one who discovered first an engine working on the dual-fuel principle [31]. The primary fuel, mostly gaseous, is injected with air into the cylinder, and then, it is further compressed [32–35]. Since the auto-ignition temperature of gaseous fuels is higher than that of liquid fuels, they cannot self-ignite, even at a high pressure. Hence, to achieve homogenous combustion, a small amount of liquid pilot fuel is required to be injected at the end of the compression stroke [36–38]. Many researchers have utilized diesel as a pilot fuel to ignite the primary air-fuel mixture [39,40]. The combustion process is similar to the SI engine [40]. Many scientists have quoted that the DFC technique improves the thermal efficiency with lesser emissions at medium to high loads [3,5,21,28,32,41]. Different feedstocks [41–46] have been used in the DFE. A summary of the results obtained from a detailed review of the literature on acetylene dual-fuel engine is shown in Table 2.2. Various graphs have been plotted to study and analyze the performance, combustion, and emissions processes based on literature review results.

2.2 PERFORMANCE ANALYSIS

Most of the researchers concluded that acetylene DFE results in lower BTE compared to baseline diesel using biodiesel and diesel as a pilot fuel [1,2,8,15,17–20,46–49], whereas Behera et al. [3,50] and Lakshmanan et al. [5,21] observed a higher thermal efficiency at high loads when using acetylene, diesel, and biodiesel combination. This may be due to the controlled and efficient combustion initiated by advanced injection techniques. Moore et al. [51] and Haragopal et al. [52] found that the performance of the DFE is superior at medium to high loads, whereas at low loads, the performance deteriorates owing to the inefficient combustion of gaseous fuels. Karim et al. [51] found that the reasons for decreasing BTE were the high self-ignition temperature of gaseous fuels and excessive pressure rise at high loads. Razavi et al. [53] suggested that using high cetane rating primary and pilot fuels can improve the performance of the DFE. Karim et al. [26] and Sahoo et al. [29,54] observed that the performance of the engine can be enhanced by increasing the quantity of pilot fuel in the DFE. Since the injected pilot fuel is very small, the performance characteristics depend on the primary gaseous fuel. Hence, the amount of primary fuel has to be appropriately metered depending on the loading conditions and gaseous fuel characteristics [55,56]. Toshio et al. [57], Mallikarjuna et al. [4,6,7], and Sahoo et al. [29,56] suggested the following methods to improve BTE: increasing excess air during mixing, reversing the direction of cooling water, and advancing the injection angle, respectively. Vijayabalan et al. [58] introduced a glow plug inside the combustion chamber, whereas Malikarajun et al. [18] increased the inlet charge temperature to increase the performance of the DFE at low loads. The variation of BTE with change in brake power is shown in Figure 2.1, which shows that the maximum efficiency is of diesel and the minimum efficiency is of the mixture of acetylene and water. The performance of all engines increases while increasing loads. This might be due to the efficient combustion achieved at peak loads. EGT rises with mounting load and was maximum at peak loading conditions. The maximum EGT is for

TABLE 2.2
Literature Review Results

Feedstock	BTE	NO_x	BC	CO	CO_2	EGT	Gas Flow Rate	PCP	Smoke (BSN)	Engine Used
Acetylene+Die Using PFI/MFI Technique [5.21]	D-28%; DFE-28%–29%	D-12.55 g/kWh DFE-(10.66–11.38) g/kWh	D-0.07g/kWh DFE-0.06g/kWh	D-0.02% vol; DFE-0.01% vol 0.56 g/kWh	D-826 g/kWh DFE 675.86 g/kWh to 778.3 g/kWh	D-444°C; DFE-388 °C, 376°C, and 380°C	110, 180, and 240 g/h	D-72 bar DFE-(69 –73 bar)	D-4; DFE increased by (5%–12%)	Single-cylinder four-stoke DI, 4.4 kW, 1500 RPM, CR-17.5 air-cooled
Acetylene+Diesel [1]	D-29% DFE-25%–27%	D-12.55 g/kWh DFE-(15.75–19.73) g/kWh	D-0.31–0.07 g/kWh DFE 0.28–0.06 g/kWh	D-0.39 g/kWh DFE 0.29–0.18 g/kWh	D-826 g/kWh DFE-802–679.75 g/kWh	D-444°C DFE-368°C, 328°C, and 301°C	0.20, 0.26, and 0.39 kg/h	D-72 bar DFE-(76–80 bar)	D-7 DFE-decreased by (7%–20%)	Single-cylinder four-stoke DI, 4.4 kW, 1500 RPM, CR-17.5 air-cooled

(Continued)

TABLE 2.2 (*Continued*)
Literature Review Results

Feedstock	BTE	NO_x	BC	CO	CO_2	EGT	Gas Flow Rate	PCP	Smoke (BSN)	Engine Used
Acetylene+Used Transformer Oil/Diesel [3,50]	D-28.6% DFE(UTO)- 28.8%–30.1% DFE(D)- 27.8%–28.5%	UTO+acetylene 10% lower than diesel+acetylene; overall increase of 23% and 9% with DFE(UTO) & DFE(D)	NA	NA	NA	D-270°C DFE(D)- 384°C 362°C, 333°C, and 299°C DFE (UTO)- 322°C, 315°C, 301°C, and 270°C	132, 198, 264, and 330 g/h	NA	UTO>D> DFE(UTO)> DFE(D)	Single-cylinder four-stoke DI, 4.4 kW, 1500 RPM, CR-17.5 air-cooled
Acetylene+ Diesel+Water [61,62]	D-29% DFE-28%; DFE+water- 25%	D-11.62g/kWh; DFE-16.93 g/kWh; DFE+water 12.09 g/kWh	D-826 g/kWh; DFE 679.75 g/kWh; DFE+-water- 600.53 g/kWh	NA	D-(0.07-0.22) g/kWh; DFE-(0.04-0.13) g/kWh; DFE+W-(0.06-0.21) g/kWh	D+444 °C; DFE- 301 °C; DFE+W- 259 °C	Gas-390 g/h Water- 200g/h, 300g/h, and 400g/h	D-71 bar; DFE-91 bar; DFE+W- 71 bar	D-4; DFE3; DFE+Water- 3.5	Single-cylinder four-stoke DI, 4.4 kW, 1500 RPM, CR-17.5 air-cooled

(*Continued*)

TABLE 2.2 (*Continued*)
Literature Review Results

Feedstock	BTE	NO_x	BC	CO	CO_2	EGT	Gas Flow Rate	PCP	Smoke (BSN)	Engine Used
Acetylene+Diesel/Linseed/Peanut [47]	D-28.74%; >DFE(-D)>DFE(-L/P) optimum at 0.5 bar for DFE(L/P)	D-543 ppm DFE(D)-615 to 625 ppm DFE(L/P)-625 to 653 ppm	D-26 ppm DFE(-D)-24 to 28 ppm DFE(-L/P)-9 to 12 ppm	D-1.16% Vol> DFE(L/P)	D-1.5% Vol> DFE(-L/P)	NA	At different pressures: 0.5 and 1 bar	D-69.9 bar DFE(D/L/P)-73 to 78 bar	NA	Single-cylinder four-stoke DI, 4.4 kW, 1500 RPM, CR-17.5 air-cooled
Acetylene+Karanja Methyl Ester [83]	D-25.31%; >DFE(-D)>DFE(K)-23.76%	D-53 ppm DFE(k)-401 to 407 ppm	D-29 ppm DFE(-K)-4 to 5 ppm	D-1.16% Vol> DFE(K)-0.02% to 0.33%	D-1.5% Vol> DFE(K)-1.4% to 1.33%	D-503 °C; DFE-582 °C	At different pressures: 0.5 and 0.8 kg/cm^2	D-69.9 bar; DFE 74 bar	NA	Single-cylinder four-stoke DI, 4.4 kW, 1500 RPM, CR-17.5 air-cooled

(*Continued*)

TABLE 2.2 (*Continued*)
Literature Review Results

Feedstock	BTE	NO_x	BC	CO	CO_2	EGT	Gas Flow Rate	PCP	Smoke (BSN)	Engine Used
Acetylene+Tamanu Methyl Ester [46]	D-29.69%; >DFE(T)-25.12%	DFE(T)-480 ppm	NA	D-1.16% Vol>DFE(T) 43% reduction using TME	NA	NA	41pm	NA	NA	Single-cylinder four-stoke DI, 4.4 kW, 1500 RPM, CR-17.5 air-cooled
Acetylene+Diesel [2]	D-25.09% at CR 20 DFE-25.72% at CR 21; both operated at 120LPH	DFE at CR 21> DFE at CR 20> D at CR 20	D at CR 20> DFE at CR 21> DFE at CR 20	D at CR 20> DFE at CR 20> DFE at CR 21 above 60% load	NA	D at CR 20> DFE at CR 20> DFE at CR 21	60 LPH, 120 LPH, 180 LPH, and 240 LPH	DFE-81.41 bar at CR 20 DFE-74.03 bar at CR 21 D-71.88 bar at CR 20	D at CR 20> DFE at CR 20> DFE at CR 21	Single-cylinder four-stoke DI, 4.4 kW, 1500 RPM, CR-15-21 air-cooled

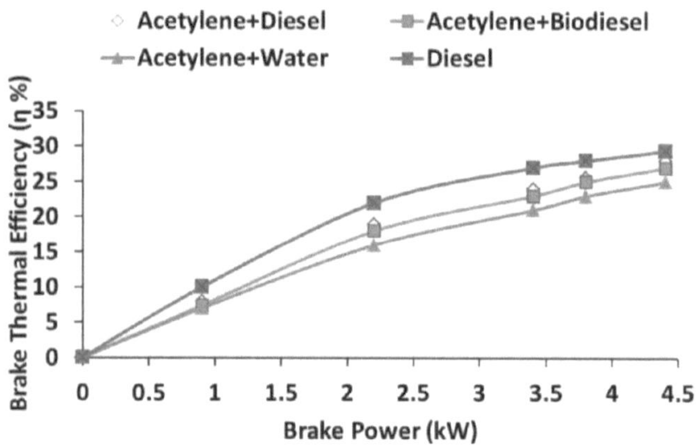

FIGURE 2.1 Variation of BTE with BP.

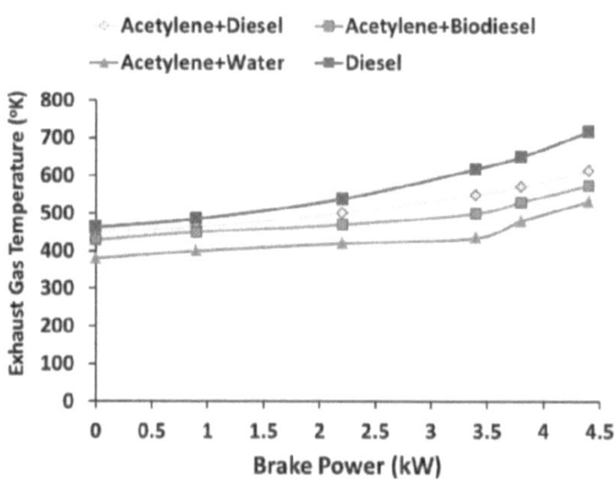

FIGURE 2.2 Variation of EGT with BP.

diesel engines, whereas the minimum is for acetylene and water engines, as shown in Figure 2.2. The introduction of acetylene gas at all loads reduces EGT. This might be due to the increase in heat transfer from gas to the wall leading to higher thermal energy loss. It has been concluded from the literature that increasing the mass flow rate of acetylene gas increases the BET and GET.

Srivastava et al. [2] analyzed the effect of the compression ratio in the acetylene dual-fuel engine and found that BTE can be increased by increasing the compression ratio. The reason might be an increase in CR; the combustion temperature may rise, which may further increase combustion efficiency, but a slight penalty in the form of NO_x can be observed. Few researchers [18,59] have studied the effects of EGR in an acetylene-fueled engine and found that BTE was improved at part load conditions. Brusca et al. [8] used optimization techniques such as GA and ANN to achieve acceptable engine performance in acetylene and alcohol DFE. A very few scientists have worked on acetylene DFE with water injection [60–62].

2.3 FUEL INJECTION SYSTEMS

Das et al. [10,12] compared different injection techniques and recommended using a manifold injection method to avoid backfiring and rapid pressure rise. Similar outcomes have been mentioned by Lakshmanan et al. [21]. Das et al. [10,12], Varde et al. [63], and many other scientists [32,33,38,41,52,64] suggested that the chances of pre-ignition and backfire in the engine cylinder can be completely eliminated by using electronically controlled injectors. Mathur et al. [10] and Soni et al. [2] suggested that a low-pressure injector should be used for intake manifold injection system, whereas Karim et al. [53,55,65] suggested that by lowering the intake temperature, a homogenous mixture can be achieved. Helium and nitrogen can be mixed during intake to achieve efficient combustion [10]. Many researchers [66–70] have stated that the reason behind inefficient combustion at medium loads is the formation of a heterogeneous mixture during the combustion process. Lakshmanan et al. [5,21] discussed the benefits of the TPI and TMI techniques, whereas Ryan et al. [71] stated the advantages of port fuel injection (PFI) injector for achieving efficient combustion. Lakshmanan et al. [5,21] found the best possible condition for intake manifold injection and suggested 10° ATDC for 90° CA. Overall, it can be concluded that three techniques have been explored to inject acetylene in CI engines; they are manifold injection (MFI), port fuel injection (PFI), and timed manifold injection (TMI). Out of these techniques, PFI and TMI techniques have been recommended by many engine scientists for efficient utilization of acetylene gas during dual-fuel combustion mode [84,85].

2.4 COMBUSTION ANALYSIS

Brusca et al. [8] analyzed the IC engine using acetylene as a primary fuel and ethanol as a secondary fuel to control detonation. The engine was modified with an ECU system with two different injectors for liquid and gaseous fuel. They also suggested operating the engine under lower-temperature combustion to increase the life expectancy of the engine. Zheng et al. and Bogin et al. [72,73] introduced a controlled auto-ignition technique to achieve stable combustion at intermediate loading conditions. Karim et al. [26,53] reported that at high intake temperatures, high compression ratios and high outputs will lead to pre-ignition and knocking, which might cause engine damage. Heat release rate (HRR) is a key factor in analyzing an IC engine's combustion efficiency, which mainly depends on the start of combustion, ignition delay and the fraction of fuel burned in the premixed mode, and differences in the combustion rates of fuels [83–85]. Figure 2.3 shows a variation of HRR of acetylene gas and diesel/biodiesel blends with the change in crank angle. The inclination of maximum HRR takes place a little earlier for oxygenated pilot fuels compared to conventional diesel, which may be due to the enhanced bulk modulus properties of oxygenated blends. The maximum heat release for neat diesel is observed at 365° CA at full load [1,5,]. Moreover, when the percentage of the oxygenated blend is

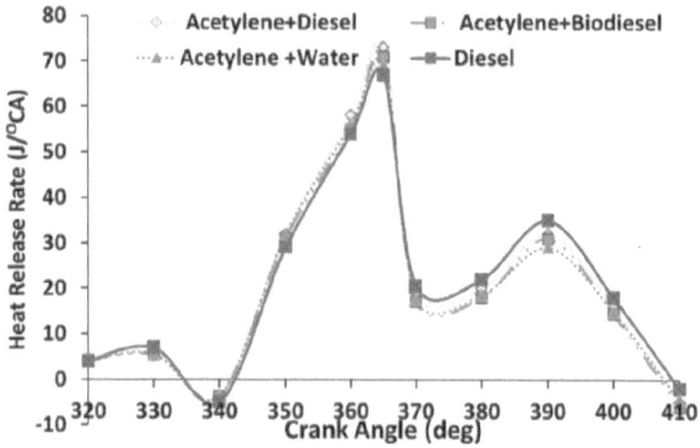

FIGURE 2.3 Variation of HRR with crank angle.

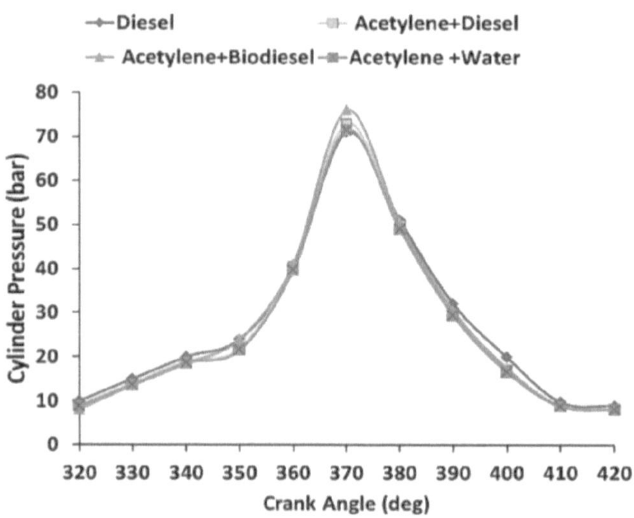

FIGURE 2.4 Variation of PCP with crank angle.

increased, the maximum HRR decreases and there is an advancement of the crank angle at which it takes place [5,19]. Most of the literature suggests that the HRR for acetylene-fueled CI engine represents a short premixed combustion phase followed by a slightly longer diffusion phase compared to baseline diesel engines [1,3]. Figure 2.4 describes the variation of cylinder pressure with crank angle. The peak pressure is nearly 72 bar for baseline diesel operation at full load. The peak pressure increases with the increase in the quantity of acetylene at full load, and it may reach 82 bar [1,5,19]. The combustion is advanced by 3 to 5°CA compared to the peak pressure incidence of the diesel engine at full load [5,50]. Most engine scientists have explained that the common reason for rising peak pressure in acetylene engines is the instantaneous occurrence of combustion of acetylene [1,5,50]. Furthermore, the acetylene-fueled engine's combustion timing varies from 1 to 3° CA earlier than that of diesel engine at the maximum load [1.19]. The reason might be the instantaneous combustion; initially, the acetylene is burnt very quickly, whereas diesel is burnt progressively at the later stage. Few researchers have also suggested that the increased BTE of DFE may be due to the higher pressure and temperature attained during the beginning of power stroke, allowing more chemical energy conversion into shaft power.

Toshio et al. [57] stated that advancing the injection angle increases the cylinder wall temperature. Liu et al. [38], Kapil et al. [48], and Sahoo et al. [29] demonstrated the effects of the mass flow rate of gaseous fuel and found that the peak pressure and exhaust gas temperature increase with increasing mass flow rate. Zheng et al. [72,74] stated that using DME, ignition can be controlled in a DFE. Karim et al. [30,65] suggested that the quantity of both the pilot fuel and gaseous fuel needs to be varied according to loading conditions for efficient combustion. Many researchers [75–78] stated that advancing the injection timing a few degrees will reduce the ignition delay of the pilot fuel. Gunea et al. [65] suggested that a gaseous fuel that has poor ignition characteristics will require a large quantity of pilot fuel, whereas readily burning fuels such as acetylene and hydrogen will require a lesser quantity to improve combustion. Karim et al. [51] stated that at peak loads, more quantity of gaseous fuel enters the engine, which produces uncontrolled combustion, leading to a high pressure rise rate ultimately leading to knock. Lakshmanan et al. [61,62] suggested that water can also be used in acetylene dual-fuel DI engines to overcome backfire and knock problems. Lakshmanan et al. [19] and Srivastava et al. [2] concluded that the maximum pressure rise rate is due to the higher flame speed of acetylene gas. Wulff et al. and Brusca et al. [8,49] suggested using alcohol to reduce the in-cylinder temperatures and pressure of the engine. A maximum HRR is observed in acetylene-diesel engines, whereas a maximum peak cylinder pressure is observed in acetylene-biodiesel engines. The reason might be the increase in the ignition delay and the rapid combustion rate of the acetylene fuel [83–85]. Acetylene induction increases diesel and biodiesel evaporation rapidly.

2.5 EMISSION ANALYSIS

Silva et al. [79] suggested that acetylene can be used as an important soot precursor and, by the use of oxygenated compounds as additives to diesel fuels, NO_x can also be reduced. Sudheesh et al. [6] and Mahla et al. [17] stated that using diethyl ether in HCCI mode results in lower NO_x and smoke emissions while increasing CO and HC emissions. Lakshmanan et al. [1,5,19], Behera et al. [3,50], Kapil et al. [48], Srivastava et al. [2], Parthasarathy et al. [46], and Raman and Kumar [85] found lower HC, CO, CO_2, and smoke emissions and higher NO_x emissions in acetylene-operated dual-fuel engines, as clearly seen from Figures 2.5 to 2.7. Lakshmanan et al. [62] and Nagarajan et al. [59] suggested that acetylene, when used with EGR and water injection, results in lower NO_x, as shown in Figure 2.7. Wulff et al. [49] and Raman and Kumar [83–85] observed a reduction in HC, CO, and NO_x emissions at peak loading conditions. Lakshmanan et al. [1,19] stated that the NO_x rate increases with an increase in peak cylinder pressure due to uncontrolled combustion during

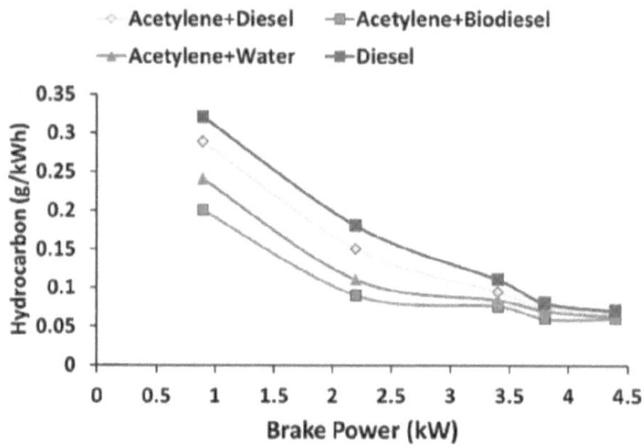

FIGURE 2.5 Variation of HC with BP.

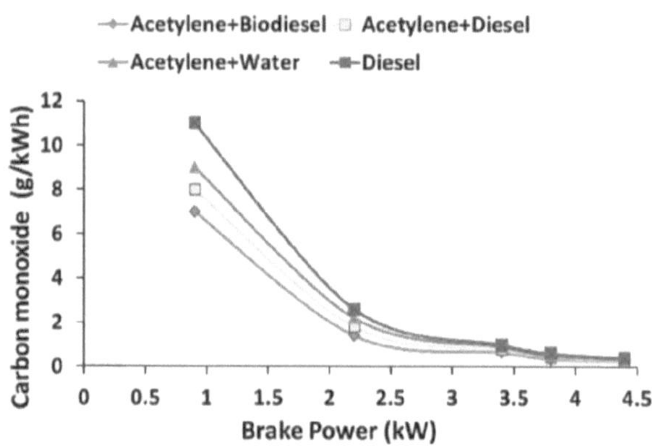

FIGURE 2.6 Variation of CO with BP.

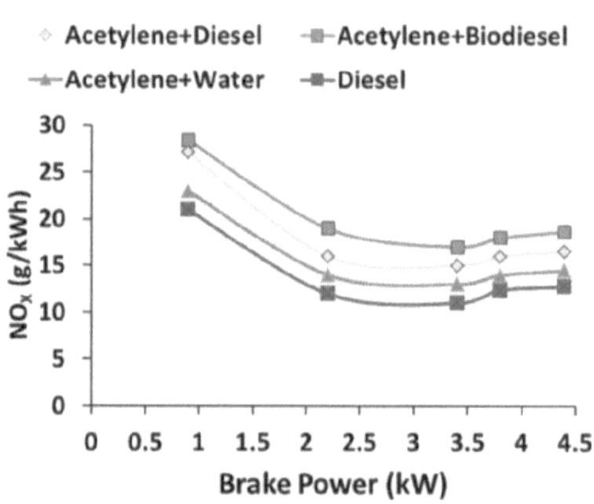

FIGURE 2.7 Variation of NO_x with BP.

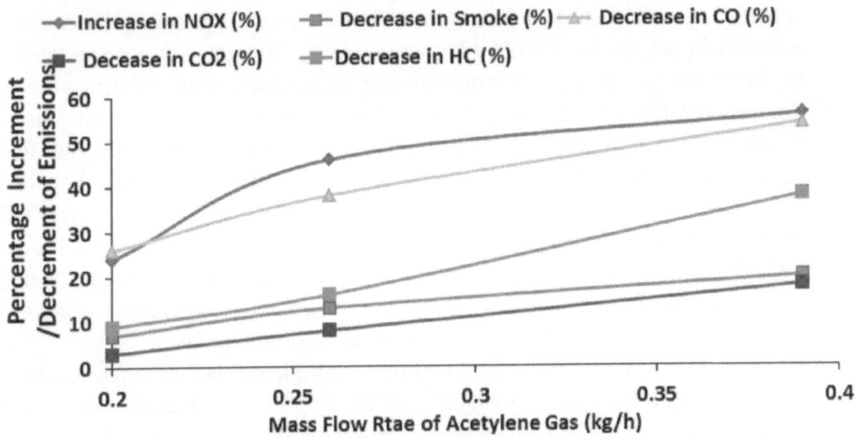

FIGURE 2.8 Percentage increment/decrement with the mass flow.

acetylene injection. Many researchers [80,81] suggested inducting gaseous fuels in the intake port of the CI engine, to lower down NO_x emission steadily. Lee et al. [82] suggested that adopting early injection technique in the DFE can reduce NO_x emissions. Basha et al. [20] performed design analysis of the modified diesel engine and studied the effect of flame velocity on the performance by CFD analysis.

Lakshmanan et al. and Behera et al. [3,5] suggested that smoke intensity and NO_x concentration can be reduced by flowing lean mixtures of gaseous fuel into the combustion chamber. The percentage increase and a decrease in emissions with acetylene induction flow rate are shown in Figure 2.8. It can be easily seen from Figures 2.5 to 2.8 that, with acetylene induction, HC and CO reduced and at the same time NO_x increases due to spontaneous combustion, which needs to be addressed seriously. The after-treatment technologies such as EGR, selective catalytic reduction (SCR), and a lean NO_x-trap (LNT) can control NO_x, or some oxygenated additives such as DEE and higher-order alcohols can also be used to get complete combustion. TPI and TMI techniques can be adapted to get proper combustion, which may reduce harmful emissions.

2.6 CONCLUSIONS

a. It can be concluded that the acetylene-diesel DFE has a higher thermal efficiency than the acetylene-biodiesel and acetylene-water engines, but a slightly smaller one than petroleum diesel. This might be because of the improved combustion characteristics of acetylene-fueled engines at intermediate to high loads.
b. The performance of the DFE depends on the quantity of acetylene as well as the quantity of the pilot fuel.
c. The BTE decreases at lower loads with a rising flow rate of acetylene gas; moreover, at high loads, the performance of the engine improves

substantially because, at low loads, the fuel-air mixture is rich and heterogeneous in nature; hence, the adequate quantity of oxygen is not available to ignite the whole mixture and thus the large quantity of unburnt gases escapes out from the exhaust, which may be the reason for low thermal efficiency at low loads. In contrast, when load increases gradually, due to homogenous charge formation, complete combustion takes place; hence, thermal efficiency improves progressively.

d. It was observed that by inducting acetylene in dual-fuel mode, smoke, soot, HC, CO, and CO_2 emissions decrease in all types of engines except for in neat diesel engines; this might be due to the heterogeneous nature of mixture formed when diesel and acetylene act as soot precursor.
e. NO_x emission rate increases in all engines at higher loads, as explained by Zeldovich mechanism; the formation of NO_x depends on the reaction temperature, availability of oxygen, and reaction period; hence, when acetylene is injected, the instantaneous combustion of gaseous fuel causes a high pressure rise rate, as observed during combustion analysis, eventually leading to a high temperature of the engine cylinder causing a higher rate of nitric oxide formation.
f. It was observed that acetylene induction decreases exhaust gas temperature at all loading conditions; this might be due to the increase in heat transfer from the gas cylinder wall to the surroundings because of the higher thermal conductivity of acetylene gas.
g. The problems of diesel engines, such as poor volatility and wall wetting, can be resolved by using acetylene DFE.
h. It is significantly easier to modify direct injection diesel engines to dual-fuel combustion modes without major modifications.

2.7 FUTURE SCOPE

The exhaustive literature suggests that the major challenges with acetylene dual-fuel engines are higher NO_x emissions, high pressure rise rate due to uncontrolled combustion, poor part-load efficiency, backfiring, and knocking at full load. So, exhaustive studies are still required to enhance acetylene dual-fuel engines' performance with lesser emissions at low loads. There is an open area to explore injection methods and to vary engine geometry to get an optimized result from an acetylene dual-fuel engine. The effects of injection pressure, timing, and angle on the engine performance have still not been fully explored. The effect of acetylene induction on engine tribology is still unexplored. The application of EGR in acetylene dual-fuel engines has not been fully explored for reducing NO_x. The use of biofuels as a pilot fuel in acetylene DFE has been limited to Karanja, linseed, peanut, and used transformer oil, etc.. The effect of using two different primary fuels simultaneously in acetylene DFE is still unexplored. The stability of acetylene with DME and DEE has not been adequately explored. A little quantum of work has been done on the effects of intake manifold temperature and exhaust gas temperature variation in the acetylene dual-fuel system. Research is underway to search for alternative fuels with better emission characteristics without compromising the performance parameters.

ABBREVIATIONS

ANN Artificial neural network
BP Brake power
BSU Bosch smoke unit
CI Compression ignition
CO Carbon monoxide
CO_2 Carbon dioxide
CR Compression ratio
D Diesel
DFE Dual-fuel engine
DI Direct injection
EGT Exhaust gas temperature
GA Genetic algorithm
HC Hydrocarbon
NO_x Oxides of nitrogen
PCP Peak cylinder pressure
PFI Port fuel injection
TMI Timed manifold injection
TPI Timed port injection
VCR Variable compression ratio

REFERENCES

[1] Lakshmanan T, Nagarajan G. Performance and emission of acetylene-aspirated diesel engine. *Jordan J Mech Ind Eng* 2009;3:125–30.
[2] Srivastava AK, Soni SL, Sharma D, Sonar D, Jain NL. Effect of compression ratio on performance, emission and combustion characteristics of diesel–acetylene-fuelled single-cylinder stationary CI engine. *Clean Technol Environ Policy* 2017;19:1361–72. doi:10.1007/s10098-017-1334-0.
[3] Behera P, Murugan S, Nagarajan G. Dual fuel operation of used transformer oil with Acetylene in a di diesel engine. *Energy Convers Manag* 2014;87:840–7. doi:10.1016/j.enconman.2014.07.034.
[4] Sudheesh K, Mallikarjuna JM. Development of an exhaust gas recirculation strategy for an acetylene-fuelled homogeneous charge compression ignition engine. *Proc Inst Mech Eng Part D J Automob Eng* 2010;224:941–52. doi:10.1243/09544070JAUTO1364.
[5] Lakshmanan T, Nagarajan G. Experimental investigation of port injection of Acetylene in diesel engine in dual fuel mode. *Fuel* 2011;90:2571–7. doi:10.1016/j.fuel.2011.03.039.
[6] Sudheesh K, Mallikarjuna JM. Diethyl ether as an ignition improver for acetylene-fuelled homogeneous charge compression ignition operation: An experimental investigation. *Int J Sustain Energy* 2015;34:561–77. doi:10.1080/14786451.2013.834338.
[7] Sudheesh K, Mallikarjuna JM. Effect of cooling water flow direction on performance of an acetylene fuelled HCCI engine. *Indian J Eng Mater Sci* 2010;17:79–85.
[8] Brusca S, Lanzafame R, Marino Cugno Garrano A, Messina M. On the possibility to run an internal combustion engine on Acetylene and alcohol. *Energy Proc* 2014;45:889–98. doi:10.1016/j.egypro.2014.01.094.
[9] Lata DB, Misra A, Medhekar S. Effect of hydrogen and LPG addition on the efficiency and emissions of a dual fuel diesel engine. *Int J Hydrogen Energy* 2012;37:6084–96. doi:10.1016/j.ijhydene.2012.01.014.

[10] Mathur HB, Das LM, Patro TN. Hydrogen-fuelled diesel engine: Performance improvement through charge dilution techniques. *Int J Hydrogen Energy* 1993;18:421–31. doi:10.1016/0360-3199(93)90221-U.

[11] Saravanan N, Nagarajan G, Sanjay G, Dhanasekaran C, Kalaiselvan KM. Combustion analysis on a DI diesel engine with hydrogen in dual fuel mode. *Fuel* 2008;87:3591–9. doi:10.1016/j.fuel.2008.07.011.

[12] Das LM. Hydrogen engine: Research and development (R&D) programmes in Indian Institute of Technology (IIT), Delhi. *Int J Hydrogen Energy* 2002;27:953–65. doi:10.1016/S0360-3199(01)00178-1.

[13] Papagiannakis RG, Hountalas DT. Combustion and exhaust emission characteristics of a dual fuel compression ignition engine operated with pilot diesel fuel and natural gas. *Energy Convers Manag* 2004. doi:10.1016/j.enconman.2004.01.013.

[14] Mustafi NN, Raine RR, Verhelst S. Combustion and emissions characteristics of a dual fuel engine operated on alternative gaseous fuels. *Fuel* 2013. doi:10.1016/j.fuel.2013.03.007.

[15] Sharma PK, Kuinkel H, Shrestha P, Poudel S. Use of acetylene as an alternative fuel in IC engine. *Eng Environ Sci* 2012.

[16] Schobert H. Production of acetylene and acetylene-based chemicals from coal. *Chem Rev* 2014. doi:10.1021/cr400276u.

[17] Mahla SK, Kumar S, Shergill H, Kumar A. Study the performance characteristics of acetylene gas in dual fuel engine with diethyl ether blends. *Engineering* 2012;3:80–3.

[18] Swami Nathan S, Mallikarjuna JM, Ramesh A. Effects of charge temperature and exhaust gas re-circulation on combustion and emission characteristics of an acetylene fuelled HCCI engine. *Fuel* 2010;89:515–21. doi:10.1016/j.fuel.2009.08.032.

[19] Lakshmanan T, Nagarajan G. Experimental investigation on dual fuel operation of Acetylene in a DI diesel engine. *Fuel Process Technol* 2010;91:496–503. doi:10.1016/j.fuproc.2009.12.010.

[20] Basha SK, Rao PS, Rajagopal K, Kotturi RK. Design and analysis of swirl in Acetylene aspirated diesel engine and its effects on performance & emissions. *Int J Latest Trends Eng Technol* 2017;8:390–9.

[21] Lakshmanan T, Nagarajan G. Experimental investigation of timed manifold injection of Acetylene in direct injection diesel engine in dual fuel mode. *Energy* 2010. doi:10.1016/j.energy.2010.03.055.

[22] Price JWH. An acetylene cylinder explosion: A most probable cause analysis. *Eng Fail Anal* 2006. doi:10.1016/j.engfailanal.2005.04.014.

[23] Lü XC, Chen W, Huang Z. A fundamental study on the control of the HCCI combustion and emissions by fuel design concept combined with controllable EGR. Part 1. The basic characteristics of HCCI combustion. *Fuel* 2005;84:1074–83. doi:10.1016/j.fuel.2004.12.014.

[24] Sher I, Levinzon-Sher D, Sher E. Miniaturization limitations of HCCI internal combustion engines. *Appl Therm Eng* 2009. doi:10.1016/j.applthermaleng.2008.03.020.

[25] Aldawood A, Mosbach S, Kraft M. HCCI combustion control using dual-fuel approach: Experimental and modeling investigations. *SAE Tech Pap Ser* 2012. doi:10.4271/2012-01-1117.

[26] Karim GA. *Dual-Fuel Diesel Engines*. CRC Press, Boca Raton, 2015.

[27] Chuahy FDF, Kokjohn SL. High efficiency dual-fuel combustion through thermochemical recovery and diesel reforming. *Appl Energy* 2017. doi:10.1016/j.apenergy.2017.03.078.

[28] Saravanan N, Nagarajan G, Kalaiselvan KM, Dhanasekaran C. An experimental investigation on hydrogen as a dual fuel for diesel engine system with exhaust gas recirculation technique. *Renew Energy* 2008. doi:10.1016/j.renene.2007.03.015.

[29] Sahoo BB, Sahoo N, Saha UK. Effect of engine parameters and type of gaseous fuel on the performance of dual-fuel gas diesel engines – A critical review. *Renew Sustain Energy Rev* 2009. doi:10.1016/j.rser.2008.08.003.

[30] Bora BJ, Saha UK. Comparative assessment of a biogas run dual fuel diesel engine with rice bran oil methyl ester, pongamia oil methyl ester and palm oil methyl ester as pilot fuels. *Renew Energy* 2015;81:490–8. doi:10.1016/j.renene.2015.03.019.
[31] Heywood JB. *Internal Combustion Engine Fundementals.* McGraw Hill Education, USA, 2018.
[32] Ma F, Wang Y, Liu H, Li Y, Wang J, Zhao S. Experimental study on thermal efficiency and emission characteristics of a lean burn hydrogen enriched natural gas engine. *Int J Hydrogen Energy* 2007. doi:10.1016/j.ijhydene.2007.07.048.
[33] Korakianitis T, Namasivayam AM, Crookes RJ. Diesel and rapeseed methyl ester (RME) pilot fuels for hydrogen and natural gas dual-fuel combustion in compression-ignition engines. *Fuel* 2011. doi:10.1016/j.fuel.2011.03.005.
[34] Ramadhas AS, Jayaraj S, Muraleedharan C. Dual fuel mode operation in diesel engines using renewable fuels: Rubber seed oil and coir-pith producer gas. *Renew Energy* 2008;33:2077–83. doi:10.1016/j.renene.2007.11.013.
[35] Yaliwal VS, Banapurmath NR, Gireesh NM, Tewari PG. Production and utilization of renewable and sustainable gaseous fuel for power generation applications: A review of literature. *Renew Sustain Energy Rev* 2014. doi:10.1016/j.rser.2014.03.043.
[36] Liu J, Yang F, Wang H, Ouyang M, Hao S. Effects of pilot fuel quantity on the emissions characteristics of a CNG/diesel dual fuel engine with optimized pilot injection timing. *Appl Energy* 2013. doi:10.1016/j.apenergy.2013.03.024.
[37] Ryu K. Effects of pilot injection pressure on the combustion and emissions characteristics in a diesel engine using biodiesel-CNG dual fuel. *Energy Convers Manag* 2013;76:506–16. doi:10.1016/j.enconman.2013.07.085.
[38] Imran S, Emberson DR, Ihracska B, Wen DS, Crookes RJ, Korakianitis T. Effect of pilot fuel quantity and type on performance and emissions of natural gas and hydrogen based combustion in a compression ignition engine. *Int J Hydrogen Energy* 2014;39:5163–75. doi:10.1016/j.ijhydene.2013.12.108.
[39] Bacha J, Freel J, Gibbs A, Gibbs L, Hemighaus G, Hoekman K, et al. Diesel fuels technical review. *Chevron* 2007. doi:10.1063/1.3575169.
[40] Galal MG, Aal MMA, El Kady MA. A comparative study between diesel and dual-fuel engines: Performance and emissions. *Combust Sci Technol* 2002. doi:10.1080/713712964.
[41] Sandalci T, Karagöz Y. Experimental investigation of the combustion characteristics, emissions and performance of hydrogen port fuel injection in a diesel engine. *Int J Hydrogen Energy* 2014. doi:10.1016/j.ijhydene.2014.09.044.
[42] Nwafor OMI. Effect of advanced injection timing on the performance of natural gas in diesel engines. *Sadhana Acad Proc Eng Sci* 2000;25:11–20. doi:10.1007/BF02703803.
[43] Yoon SH, Lee CS. Experimental investigation on the combustion and exhaust emission characteristics of biogas-biodiesel dual-fuel combustion in a CI engine. *Fuel Process Technol* 2011. doi:10.1016/j.fuproc.2010.12.021.
[44] Barik D, Murugan S. Experimental investigation on the behavior of a DI diesel engine fueled with raw biogas–diesel dual fuel at different injection timing. *J Energy Inst* 2016. doi:10.1016/j.joei.2015.03.002.
[45] Selim MYE, Radwan MS, Saleh HE. Improving the performance of dual fuel engines running on natural gas/LPG by using pilot fuel derived from jojoba seeds. *Renew Energy* 2008. doi:10.1016/j.renene.2007.07.015.
[46] Parthasarathy M, Isaac Joshua Ramesh Lalvani J, Muhilan P, Dhinesh B, Annamalai K. Experimental study of acetylene enriched air in di diesel engine powered by biodiesel-diesel blends. *J Chem Pharm Sci* 2014.
[47] Valmiki S. Dual fuel operation of performance and emission characteristics of linseed biodiesel using acetylene gas. *Int Res J Eng Tech* 2016:114–20.

[48] Choudhary KD, Nayyar A. Optimization of induction flow rate of acetylene in the C. I. engine operated on duel fuel mode. *Int J Emerg Technol Adv Eng* 2013;3(12):297–302.
[49] Wulff, JW, Maynard H, Sunggyu L. Dual fuel composition including Acetylene for use with diesel and other internal combustion engines. United States Patent 2001.
[50] Behera P, Murugan S. Combustion, performance and emission parameters of used transformer oil and its diesel blends in a di diesel engine. *Fuel* 2013;104:147–54. doi:10.1016/j.fuel.2012.09.077.
[51] Karim GA, Klat SR, Moore NPW. Knock in dual-fuel engines. *Proc Inst Mech Eng* 1966. doi:10.1002/poc.1492.
[52] Haragopala Rao B, Shrivastava KN, Bhakta HN. Hydrogen for dual fuel engine operation. *Int J Hydrogen Energy* 1983. doi:10.1016/0360-3199(83)90054-X.
[53] Karim GA. The dual fuel engine of the compression ignition type – Prospects, problems and solutions – A review. *Nat Das Dual Fuel Engine* 1983. doi:10.4271/831073.
[54] Abd Alla GH, Soliman HA, Badr OA, Abd Rabbo MF. Effect of pilot fuel quantity on the performance of a dual fuel engine. *Energy Convers Manag* 2000. doi:10.1016/S0196-8904(99)00124-7.
[55] Liu Z, Karim GA. Simulation of combustion processes in gas-fuelled diesel engines. *Proc Inst Mech Eng Part A J Power Energy* 1997. doi:10.1243/0957650971537079.
[56] Sahoo BB, Sahoo N, Saha UK. Effect of H2:CO ratio in syngas on the performance of a dual fuel diesel engine operation. *Appl Therm Eng* 2012;49:139–46. doi:10.1016/j.applthermaleng.2011.08.021.
[57] Shudo T, Yamada H. Hydrogen as an ignition-controlling agent for HCCI combustion engine by suppressing the low-temperature oxidation. *Int J Hydrogen Energy* 2007. doi:10.1016/j.ijhydene.2006.12.002.
[58] Vijayabalan P, Nagarajan G. Performance, emission and combustion of lpg diesel dual fuel. *Jordan J Mech Ind Eng* 2009;3(2):105–110.
[59] Lakshmanan T, Nagarajan G. Study on using Acetylene in dual fuel mode with exhaust gas recirculation. *Energy* 2011. doi:10.1016/j.energy.2011.03.061.
[60] Brusca S, Lanzafame R. Evaluation of the effects of water injection in a single cylinder CFR cetane engine. *SAE Tech Pap Ser* 2010. doi:10.4271/2001-01-2012.
[61] Lakshmanan T, Ahmed AK, Nagarajan G. Effect of water injection in acetylene-diesel dual fuel DI diesel engine. Proceedings of the ASME 2012 Internal Combustion Engine Division Fall Technical Conference. ASME 2012 Internal Combustion Engine Division Fall Technical Conference. Vancouver, BC, Canada. September 23–26, 2012. pp. 177–181. ASME. doi:10.1115/ICEF2012-92145.
[62] Varde KS, Frame GM. A study of combustion and engine performance using electronic hydrogen fuel injection. *Int J Hydrogen Energy* 1984. doi:10.1016/0360-3199(84)90085-5.
[63] Bendu H, Murugan S. Homogeneous charge compression ignition (HCCI) combustion: Mixture preparation and control strategies in diesel engines. *Renew Sustain Energy Rev* 2014. doi:10.1016/j.rser.2014.07.019.
[64] Gunea C, Razavi MRM, Karim GA. The effects of pilot fuel quality on dual fuel engine ignition delay. *SAE Tech Pap Ser*, 2010. doi:10.4271/982453.
[65] Sharma TK, Rao GAP, Murthy KM. Homogeneous charge compression ignition (HCCI) engines: A review. *Arch Comput Methods Eng* 2016. doi:10.1007/s11831-015-9153-0.
[66] Ganesh D, Nagarajan G, Mohamed Ibrahim M. Study of performance, combustion and emission characteristics of diesel homogeneous charge compression ignition (HCCI) combustion with external mixture formation. *Fuel* 2008. doi:10.1016/j.fuel.2008.06.010.
[67] Vinayagam N, Nagarajan G. Experimental study of performance and emission characteristics of DEE-assisted minimally processed ethanol fuelled HCCI engine. *Int J Automot Technol* 2014. doi:10.1007/s12239-014-0054-2.

[68] Bhaskar K, Nagarajan G, Sampath S. The effects of premixed ratios on the performance and emission of PPCCI combustion in a single cylinder diesel engine. *Int J Green Energy* 2013. doi:10.1080/15435075.2011.647364.
[69] Ganesh D, Nagarajan G. Homogeneous charge compression ignition (HCCI) combustion of diesel fuel with external mixture formation. *Energy* 2009. doi:10.1016/j.energy.2009.09.005.
[70] Ryan TW, Callahan TJ. Homogeneous charge compression ignition of diesel fuel. *SAE Tech Pap Ser*, 2010. doi:10.4271/961160.
[71] Zheng Z, Yao M, Chen Z, Zhang B. Experimental study on HCCI combustion of dimethyl ether(DME)/methanol dual fuel. *SAE Tech Pap Ser*, 2010. doi:10.4271/2004-01-2993.
[72] Bogin GE, Mack JH, Dibble RW. Homogeneous charge compression ignition (HCCI) engine. *SAE Int J Fuels Lubr* 2010. doi:10.4271/2009-01-1805.
[73] Chen Z, Yao M, Zheng Z, Zhang Q. Experimental and numerical study of methanol/dimethyl ether dual-fuel compound combustion. *Energy Fuels* 2009. doi:10.1021/ef8010542.
[74] Agarwal AK, Srivastava DK, Dhar A, Maurya RK, Shukla PC, Singh AP. Effect of fuel injection timing and pressure on combustion, emissions and performance characteristics of a single cylinder diesel engine. *Fuel* 2013. doi:10.1016/j.fuel.2013.03.016.
[75] Kannan GR, Anand R. Effect of injection pressure and injection timing on DI diesel engine fuelled with biodiesel from waste cooking oil. *Biomass Bioenergy* 2012. doi:10.1016/j.biombioe.2012.08.006.
[76] Gnanasekaran S, Saravanan N, Ilangkumaran M. Influence of injection timing on performance, emission and combustion characteristics of a DI diesel engine running on fish oil biodiesel. *Energy* 2016. doi:10.1016/j.energy.2016.10.039.
[77] Ganapathy T, Gakkhar RP, Murugesan K. Influence of injection timing on performance, combustion and emission characteristics of Jatropha biodiesel engine. *Appl Energy* 2011. doi:10.1016/j.apenergy.2011.05.016.
[78] Abián M, Silva SL, Millera Á, Bilbao R, Alzueta MU. Effect of operating conditions on NO reduction by acetylene-ethanol mixtures. *Fuel Process Technol* 2010. doi:10.1016/j.fuproc.2010.03.034.
[79] Uludogan A, Xin J, Reitz RD. Exploring the use of multiple injectors and split injection to reduce DI diesel engine emissions. *SAE Tech Pap Ser* 2010. doi:10.4271/962058.
[80] Hess MA, Haas MJ, Foglia TA. Attempts to reduce NO_x exhaust emissions by using reformulated biodiesel. *Fuel Process Technol* 2007. doi:10.1016/j.fuproc.2007.02.001.
[81] Lee J, Chu S, Cha J, Choi H, Min K. Effect of the diesel injection strategy on the combustion and emissions of propane/diesel dual fuel premixed charge compression ignition engines. *Energy* 2015. doi:10.1016/j.energy.2015.09.032.
[82] Sudheer M, Nagaraj R. Experimental investigation of performance, emission and combustion characteristics on ci engine using karanja biodiesel with acetylene (welding gas) blend. *Int J Sci Res Dev* 2016;4(6):4. ISSN 2321-0613.
[83] Raman R, Kumar N. Experimental investigation to analyze the effect of induction length of diesel-acetylene dual fuel engine. *Energy Sources A* 2019. doi:10.1080/15567036.2019.1663314.
[84] Raman R, Kumar N. The utilization of n-butanol78ih/diesel blends in acetylene dual fuel engine. *Energy Reports* 2019. doi:10.1016/j.egyr.2019.08.005.
[85] Raman R, Kumar N. Experimental studies to evaluate the combustion, performance and emission characteristics of acetylene fuelled CI engine. *Int. J. Ambient Energy* 2019; 1–28. doi:10.1080/01430750.2019.1709896.

3 Investigation on Stand-Alone Solar Energy Conversion System with Artificial Intelligence Techniques

Neeraj Tiwari
Poornima University

Suman Dutta and Alok Kumar Singh
Dr. K. N. Modi University

CONTENTS

3.1 INTRODUCTION

The solar photovoltaic (SPV) power system is widely used due to its cost-effectiveness and high efficiency [1]. Renewable energy sources (RES) are considered as one of

DOI: 10.1201/9781003272717-3

the most promising resources nowadays because of their cleanliness, abundance, and environmental friendliness, compared with the energy sources such as thermal, oil, natural gas, and fossil fuels [2]. Despite their advantages, the output active power P from the solar power system varies according to the solar irradiance and operation temperature T, especially under rapidly changing partial shading condition (PSC) due to the non-linear characteristic of photovoltaic (PV) cells [3]. The complex relationship between the power output and input parameters of PV cells results in unsatisfactory power extraction [4]. To alleviate the aforementioned limitations, the MPPT becomes the focus of research for the improvement of the efficiency η of solar power systems and ensuring that the operation point is always at maximum [5]. The peak uniform conditions without PSC can be tracked effectively by using conventional hill climbing (HC) MPPT techniques, i.e., P&O and INC techniques [6]. However, the power output from solar power systems generates multiple peaks under PSC, including one global MPP (GMPP) and many other local peaks. This complicates the HC MPPT technique to search for the real maximum [7]. Hence, the MPPT evolves into an algorithm based on evolutionary, heuristic, and meta-heuristic techniques. It is designated to track global peak instead of local peaks since conventional HC MPPT techniques fail to track global peak under PSC and rapidly changing solar irradiance [8]. Apart from electronically implemented MPPTs, there are other techniques to improve solar energy efficiency, such as integrated soft computing weather forecast and adjustment of the tilting angle of solar panel to track the sun direction [9]. We only focus on the artificial intelligence (AI)-based MPPT techniques for DC-to-DC converter in the solar photovoltaic (SPV) power system.

The integration of various AI optimization techniques with MPPT is aimed to resolve and rectify the following limitations of a conventional HC MPPT:

a. Self-learning capabilities.
b. More oscillation at maximum power point.
c. Very slow transient response.
d. Inability to find GMPP [10].

In general, the existing AI-based MPPT techniques utilize the sensory information including solar irradiance Ee, input voltage of solar power system $V_I PV$, and input current $I_I PV$ measurements to predict and estimate the GMPP throughout the non-linear P-V curve. The integration of AI in MPPT accelerates the convergence speed and transient response because of its complex, robust, self-learning, and digitalized system. MPPT techniques are categorized into two major groups: conventional HC MPPT and AI-based MPPT [11]. AI-based MPPT is known as computational intelligence (CI)-based MPPT, soft computing MPPT, modern MPPT, or bio-inspired MPPT. It mainly comprises of fuzzy logic control (FLC), cuckoo search (CS), firefly algorithm (FA), and hybrid algorithms. Conventional HC MPPT techniques consist of P&O, IC, HC, constant voltage, scanning-tracking of current-voltage (I-V) curve, Fibonacci searching, global MPPT (GMPPT) segmentation searching, and extremum seeking control. There are various sources of comparative literature review for all types of MPPT. The existing literature only covers AI-based and hybrid MPPT techniques. There are very limited comparative studies, specifically in AI-based

MPPT techniques [11–13]. The objective of this paper is to ensure maximum power by coordinating appropriate control strategy with sources and presenting the comparative results at different inputs and at different load conditions. The Simulink model of the MPPT developed by using two algorithms, artificial intelligence-based MPPT and INC-based MPPT for the PV system.

3.2 MATHEMATICAL MODELING AND SIMULINK MODEL

The design and implementation of various components of the device proposed, modeling of PV array, and the study of power electronics interfacing devices such as DC–AC converter and DC–DC converter are given in this section. It gives an overall idea about the proposed system. The system is developed using MATLAB/Simulink.

3.2.1 An Equivalent PV System Model

The solar cell plays a role in the conversion of sun rays of particular wavelength into electrical energy. This module comprises of various solar cells connected in series and parallel. The circuit diagram of a solar cell is presented in Figure 3.1. The electrical characteristics of a cell with load line decide the operating condition of an SPV plant, where the load line intersect the I–V curve. In this section, the mathematical model of solar cells is presented.

The voltage–current relationship can be written as:

$$I = I_L - I_D = I_L - I_s\left\{e\frac{q(V+\mathrm{Re})}{AKT} - 1\right\} - \frac{V+IR_e}{R_{sh}} \tag{1.1}$$

It is possible to enumerate I_L:

$$I_L = \frac{\phi}{\phi_{\text{ref}}}\left[I_{L,\text{ref}} + \mu_{sc}\left(T_C - T_{c,\text{ref}}\right)\right] \tag{1.2}$$

The saturation current I_s can be expressed at the reference condition as:

$$I_s = I_{C,\text{ref}}\left(\frac{T_{C,\text{ref}}+273}{T_C+273}\right)^3 \exp\left[\frac{e_{\text{gap}}N_S}{q_{\text{ref}}}\left(1-\frac{T_{C,\text{ref}}+273}{T_C+273}\right)\right] \tag{1.3}$$

The $I_{s,\text{ ref}}$ can be expressed as:

$$I_{s,\text{ref}} = I_{L,\text{ref}}\exp\left(-\frac{V_{oc,\text{ref}}}{\alpha_{\text{ref}}}\right) \tag{1.4}$$

The value of open-circuit voltage at reference condition is given by the manufacturer.

The value of α_{ref} can be calculated by:

$$\alpha_{\text{ref}} = \frac{2V_{mpp,\text{ref}} - V_{oc,\text{ref}}}{\dfrac{I_{sc,\text{ref}}}{I_{sc,\text{ref}} - I_{mpp,\text{ref}}} + In\left(1-\dfrac{I_{mpp,\text{ref}}}{I_{sc,\text{ref}}}\right)} \tag{1.5}$$

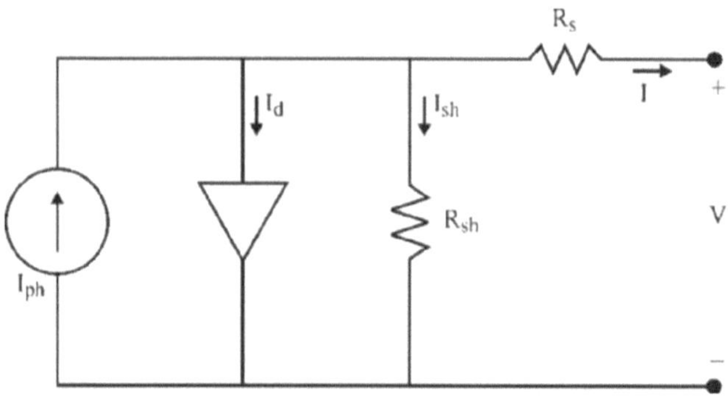

FIGURE 3.1 Equivalent circuit of a solar PV cell.

Here, α is a function of temperature. The value of α can be calculated by the following equation:

$$\alpha = \frac{T_c + 273}{T_{c,\text{ref}} + 273}\alpha_{\text{ref}} \tag{1.6}$$

The value of series resistance is provided by some manufacturers. To estimate the value of R_s, the following equation can be used:

$$R_s = \frac{\alpha_{\text{ref}} In\left(\frac{I_{mpp,\text{ref}}}{I_{sc,\text{ref}} - I_{mpp,\text{ref}}}\right) + V_{oc,\text{ref}} - V_{mpp,\text{ref}}}{I_{mpp,\text{ref}}} \tag{1.7}$$

After the study of the PV module, the effect of temperature on I–V characteristic can be seen. As the temperature value increases, the output open-circuit voltage decreases. Many researchers have proposed different methods for controlling the temperature of the module, i.e., air cooling method and water flow cooling method (Table 3.1). The temperature of a PV module varies when there is a change in irradiance and its output current and voltage, and the equation can be expressed as:

$$C_{pv}\frac{dT_c}{dt} = k_{a,PV}\phi - \frac{VI}{A} - k_{\text{loss}}\left(T_c - T_a\right) \tag{1.8}$$

3.2.2 Artificial Neural Network-Based MPPT for Solar PV System

The artificial neural network (ANN) technique provides better results as compared to other algorithms. The implementation of ANN in MPPT charge controller is done for tracking the maximum power point on the PV curve. The basic structure of an ANN consists of a layer system; usually, three layers are used. The first layer is the input layer, the second layer is called hidden layer, and the third layer is the output layer. Figure 3.2 shows the architecture of ANN layers.

TABLE 3.1
SPV Parameters Used in Simulation

Total capacity of the PV system	2558 W
Maximum power of PV cell (W)	213.15 W
Maximum power point voltage (Vmpp)	29 V
Maximum power point current(A)	7.35 A
PV system open circuit voltage (VQC)	36.3 V
PV system short circuit current(A)	7.84 A

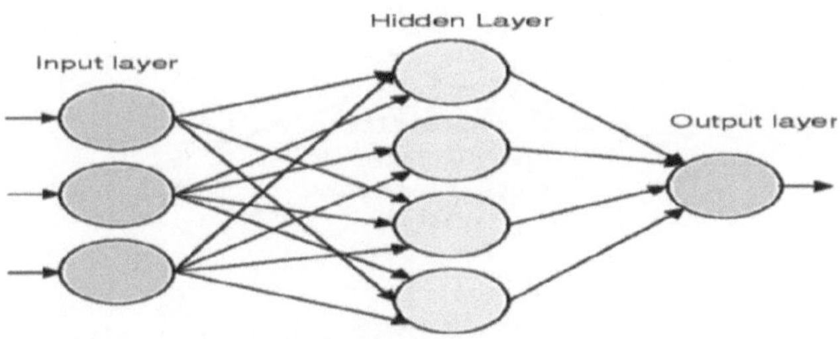

FIGURE 3.2 Basic architecture of an ANN.

In the simulation results, implementation of ANN, to track the MPP in MPPT charge controller. It gives accurate and fast response for different environmental conditions. The Levenberg-Marquardt algorithm is used to train the neural network for the MPPT of solar PV systems. The data used for the training of the ANN technique used in the MPPT are gained from the PV module. A flowchart of the training procedure of ANN is shown in Figure 3.3.

3.2.3 Simulink Model

The neural network for a PV system consists of two input variables: voltage and current. The hidden layer consists of 15 neurons. An algorithm proposed by Levenberg-Marquardt is used in this training of the neural network for the MPPT. The proposed MPPT scheme is shown in Figure 3.4 while, Figure 3.5 shows the architecture of the Levenberg-Marquardt ANN.

3.3 SIMULATION RESULTS AND DISCUSSION

The developed Simulink model system consists of a 2.5 kW standalone solar photovoltaic system with MPPT controllers. The Simscape toolbox is used for the model development. The proposed system has been verified under different input/output conditions. In this section, a comparative study of the conventional incremental conductance MPPT controller and ANN-based Levenberg algorithm is presented. Simulations are carried out in different cases, which are displayed and discussed in the following sections.

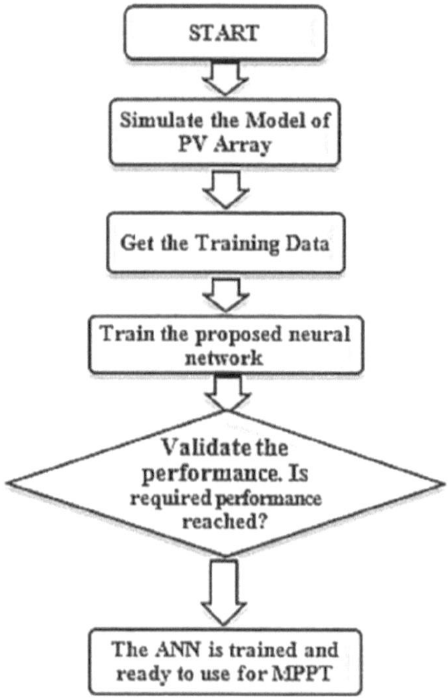

FIGURE 3.3 Training procedure of an ANN for solar PV systems.

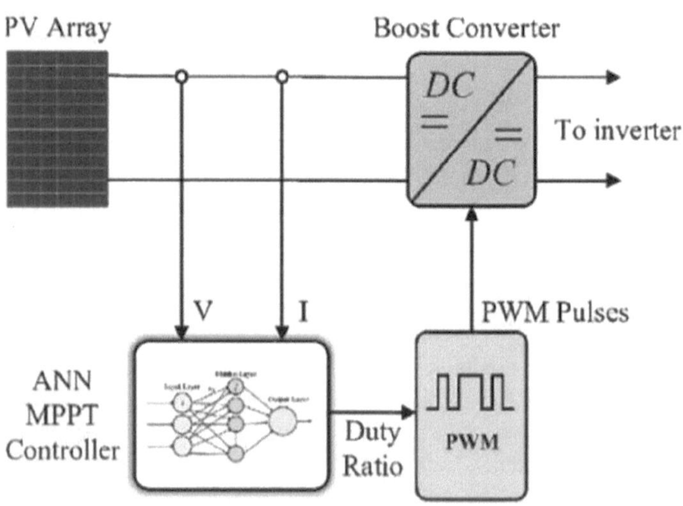

FIGURE 3.4 The proposed MPPT controller.

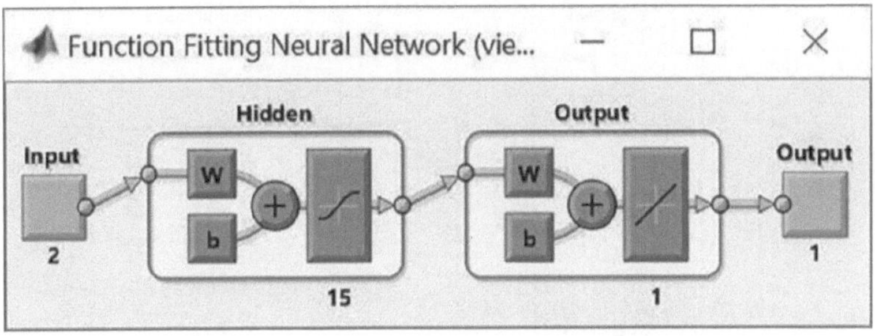

FIGURE 3.5 Architecture of Levenberg-Marquardt ANN.

3.3.1 Simulation Result of Levenberg-Based ANN MPPT Controller at Irradiance Step Change from 1000–800 to 600–400 W/m² with Resistive Load

This simulation study shows the proposed system performance with the Levenberg-based ANN MPPT controller at the step change from 1000–800 to 600–400 W/m^2 with a resistive load of 2.5 kW. In this case, the model is simulated for 4 seconds. Simulation results at different stages are shown in the following subsections.

3.3.1.1 PV Array Results

Figure 3.6a shows the input irradiance with varying irradiance levels while the temperature is kept constant; i.e., at $t=0$, the irradiance is 1000 W/m^2; it is changed to 800 W/m^2 at $t=1$ second; it is again changed to 600 W/m^2 at $t=2$ seconds; and at $t=3$ seconds, it is changed to 400 W/m^2 with the temperature being maintained at 25°C throughout the whole simulation process, as shown in Figure 3.6b. The output voltage of the PV system is nearly 170 V during the total simulation time and slightly changes at time $t=1$, $t=2$, and $t=3$ seconds when the step is changed from 1000–800 to 600–400 W/m^2, as presented in Figure 3.6c. Figure 3.6d shows the PV system output current that changes from 15–12 to 9–6 A during $t=0$–4 seconds.

Figure 3.6f presents the output power at different levels of the system starting from 2550–2050 to 1550–1000 W. When the step change in irradiance level is from "1000–800 to 600–400 W/m^2" at $t=3$ seconds, the PV output power reaches 1000 W.

3.3.1.2 Boost Converter Results

In this section, boost converter results are displayed at the step change in irradiance from "1000–800 to 600–400 W/m^2" with a resistive load.

Figure 3.7a shows the voltage output of boost converter, which is kept constant at 500 V, and Figure 3.7b shows the current generated by the Levenberg-based ANN MPPT controller at the step decrease in irradiance from "1000–800 to 600–400 W/m^2" with a resistive load of 2.5 kW. Figure 3.7c shows the duty cycle, and Figure 3.7d shows the switching pulse of the boost converter. This duty ratio is the input to the pulse generator, which generates PWM pulses of switching frequency 5 kHz, and the duty ratio changes with the irradiance.

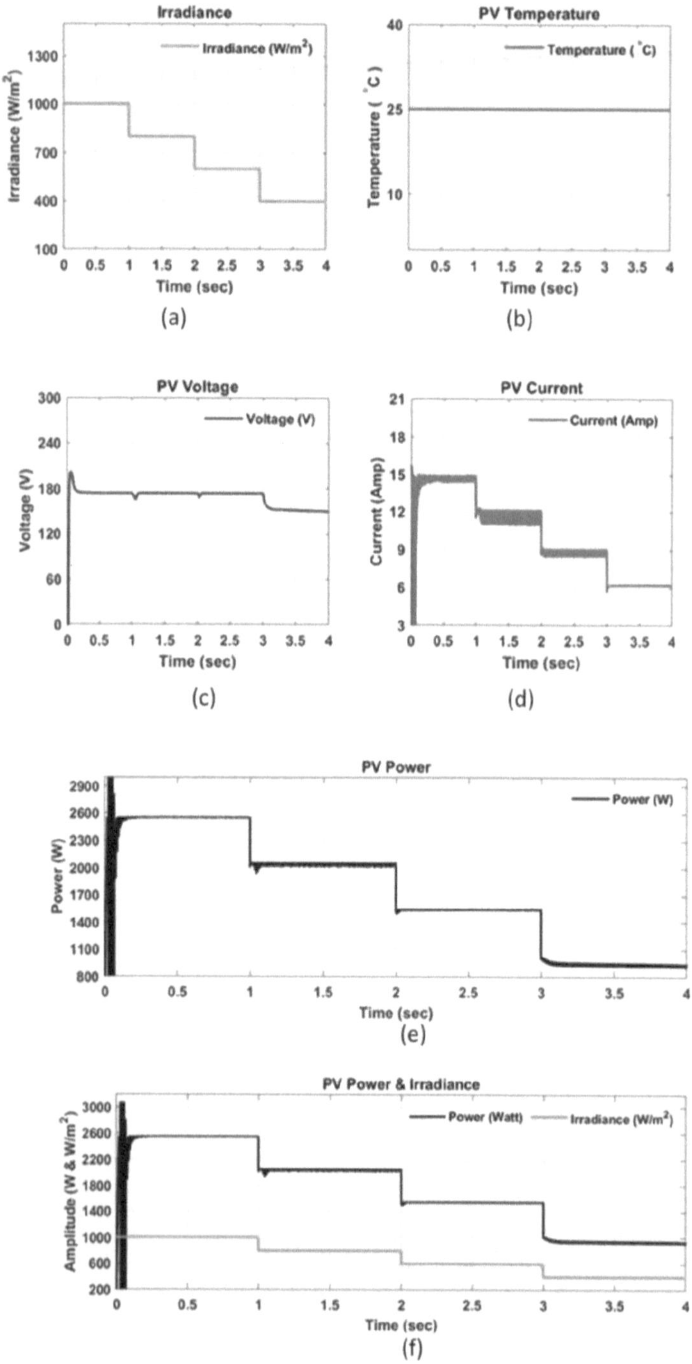

FIGURE 3.6 PV array simulation output at the step change in irradiance from 1000–800 to 600–400 W/m² with a resistive load: (a) irradiance, (b) temperature, (c) PV voltage, (d) PV current, (e) PV actual output power, and (f) PV output power with irradiance.

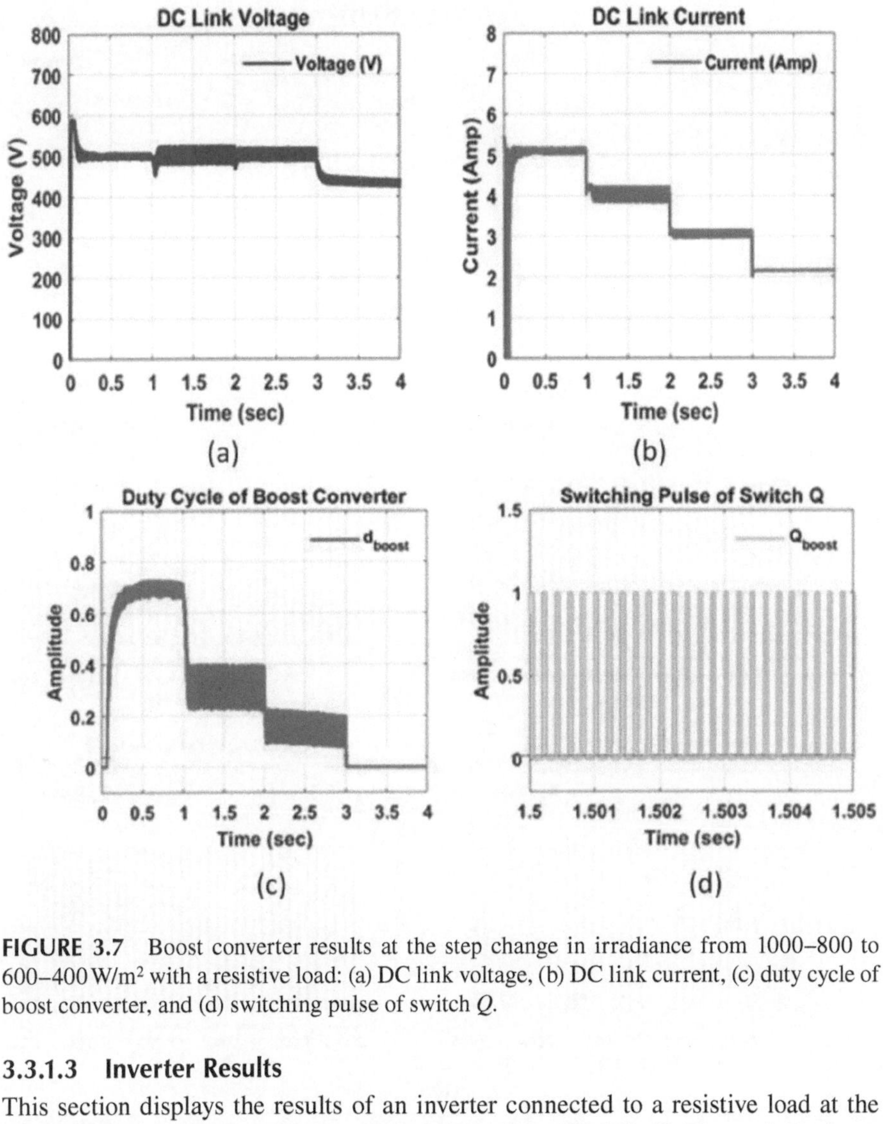

FIGURE 3.7 Boost converter results at the step change in irradiance from 1000–800 to 600–400 W/m^2 with a resistive load: (a) DC link voltage, (b) DC link current, (c) duty cycle of boost converter, and (d) switching pulse of switch *Q*.

3.3.1.3 Inverter Results

This section displays the results of an inverter connected to a resistive load at the step change in irradiance from "1000–800 to 600–400 W/m^2". Figure 3.8a shows the duty cycle of the inverter. Figure 3.8b shows the switching pulses of different switches for an inverter at the change in irradiance from "1000–800 to 600–400 W/m^2" with a resistive load. The duty ratio is slightly changed at different time levels $t=1$ than $t=2$ after that $t=3$ seconds when the irradiance changes from "1000–800 to 600–400 W/m^2".

3.3.1.4 Load Side Results

In this section, the load side simulation results are discussed at the step change in irradiance from "1000–800 to 600–400 W/m^2" with a resistive load. Figure 3.9a and b shows the output voltage and output current waveforms of the load is change

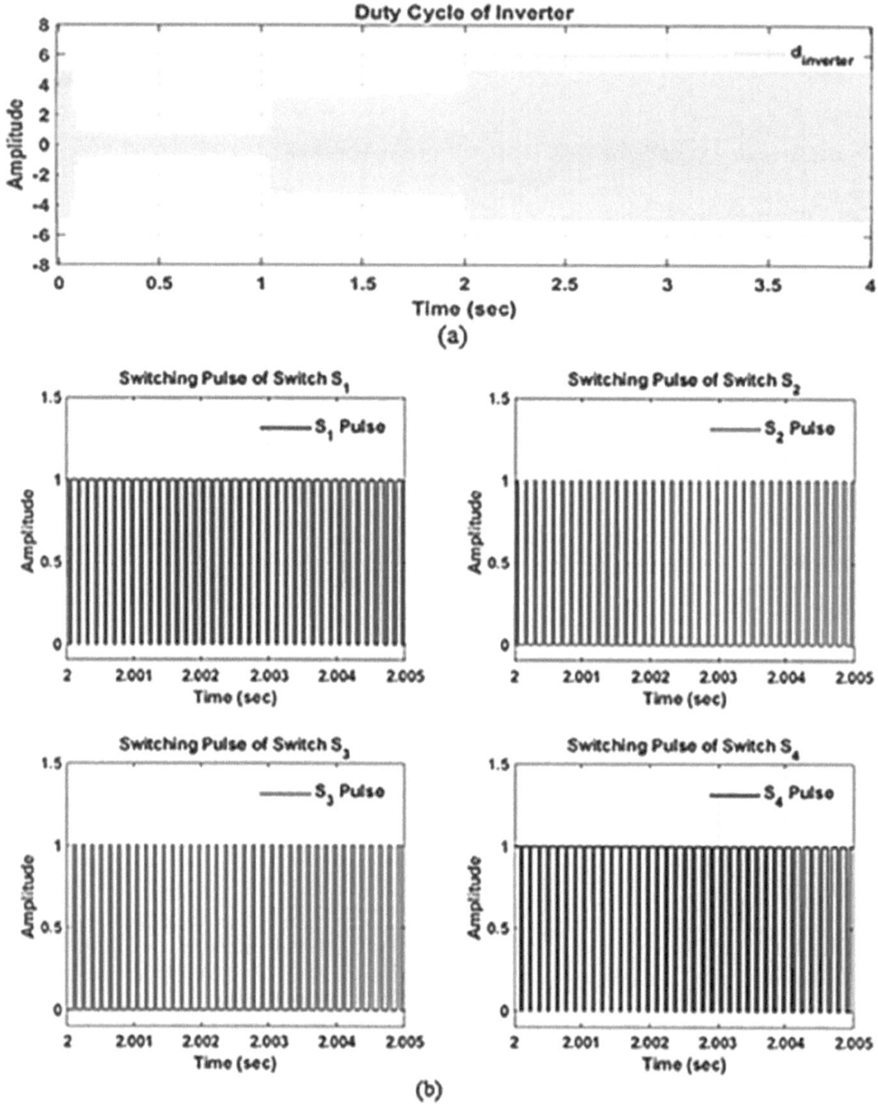

FIGURE 3.8 Inverter results at the step change in irradiance from 1000–800 to 600–400 W/m^2 with a resistive load: (a) duty cycle of inverter and (b) switching pulses of switches S_1, S_2, S_3, and S_4.

according to irradiance. Figure 3.9c from 300 to 130 V, respectively, change in irradiance from 1000 to 400 W/m^2 in steps. Figure 3.9d shows the change in PV power output with change in irridiance. The PV power output is 2400 W at 1000 W/m^2 reduces to 2000 W at 800 W/m^2 and 1400 W at 600 W/m^2 and finally reach at 1000 W at 400 W/m^2. We can see in Figure 3.9e that the PV output voltage and load voltage are nearly equal, so we can say the system is well regulated and stable.

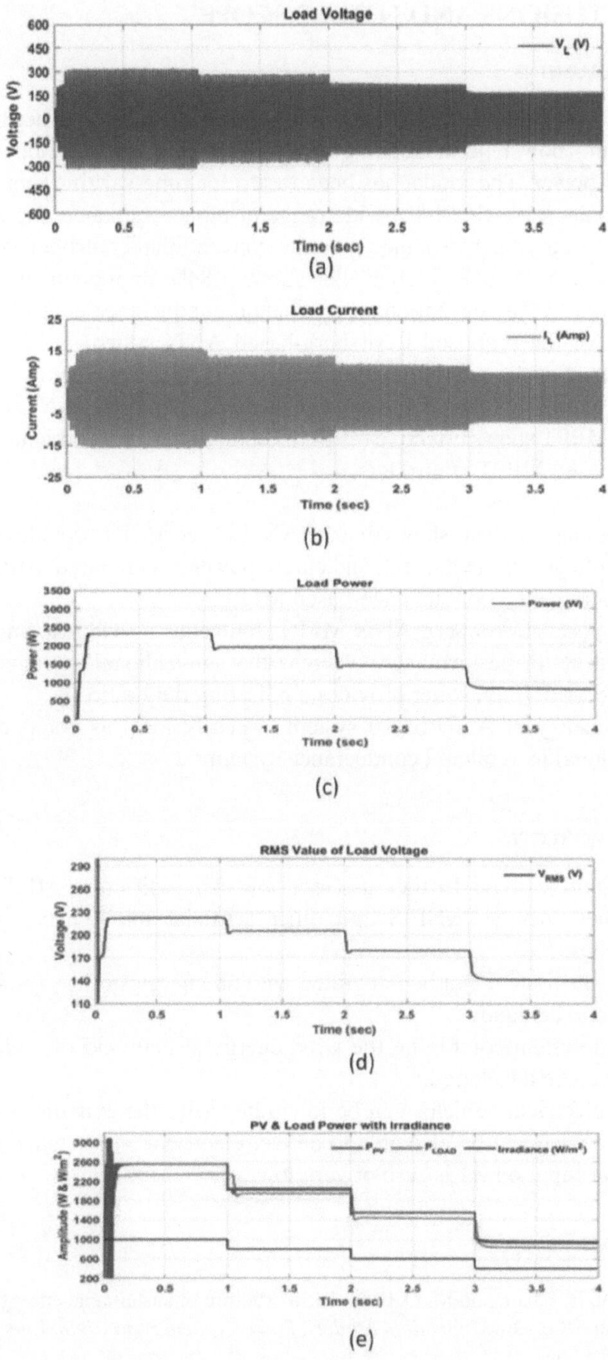

FIGURE 3.9 Load side results at the step change in irradiance from 1000–800 to 600–400 W/m^2 with a resistive load: (a) load voltage, (b) load current, (c) load power, (d) RMS value of load voltage, and (e) PV power and load power with irradiance.

3.4 CONCLUSIONS AND FUTURE SCOPE

3.4.1 Conclusions

In this paper, a standalone solar PV system with artificial neural network (ANN)-based maximum power point tracking (MPPT) technique is developed to achieve the maximum power. The model has been tested for time-varying input conditions. Simulation results show that for a wide range of input irradiance, the ANN MPPT controller shows better performance than the conventional controller. The proposed Levenberg-based ANN MPPT controller shows 1.84% improvement in PV power output compared to the conventional incremental conductance controller. The average efficiency of the proposed Levenberg-based ANN system is 99.57%, and the average efficiency of the conventional incremental conductance controller is 97.73%. From the simulation results of the system, it is clear that the ANN-based MPPT algorithm for MPPT shows effective results as compared to conventional incremental conductance-based MPPT controllers.

- The simulated analysis shows that the ANN-based MPPT provides improved output voltage, output current, and output power as compared to the conventional incremental conductance-based MPPT.
- The proposed Levenberg ANN MPPT controller provides a smooth duty ratio and better performance, whereas the conventional incremental conductance MPPT controller provides a pulsating duty ratio.
- The efficiency of ANN-based system is nearly ideal as compared to the conventional incremental conductance system.

3.4.2 Future Scope

- The system can be further assessed by using different MPPT control techniques such as ANFIS and other artificial intelligence (AI)-based approaches.
- Further, the system can be integrated with an energy storage system to fulfill the load demand.
- A hybrid system combining the wind energy system and the solar energy system can be developed.
- Plug in electrical vehicles can be integrated with the existing system as a storage device and be also able to deliver electrical power when the available solar input power is zero or very low.

REFERENCES

[1] S. Kanwal, B. Khan, and M. Q. Rauf, Infrastructure of sustainable energy development in Pakistan: A review. *Journal of Modern Power Systems and Clean Energy*, 8, 2, 206–218, 2020.

[2] X. Chen, M. B. Mcelroy, Q. Wu et al., Transition towards higher penetration of renewables: An overview of interlinked technical, environmental and socio-economic challenges. *Journal of Modern Power Systems and Clean Energy*, 7, 1, 1–8, 2019.

[3] S. Bhattacharyya, D. S. K. Patnam, S. Samanta et al., Steady output and fast tracking MPPT (SOFT MPPT) for P&O and InC algorithm. *IEEE Transactions on Sustainable Energy*. doi: 10.1109/TSTE. 2020.2991768.

[4] Y. Yang and H. Wen, Adaptive perturb and observe maximum power point tracking with current predictive and decoupled power control for grid-connected photovoltaic inverters. *Journal of Modern Power Systems and Clean Energy*, 7, 2, 422–432, 2019.

[5] K. Y. Yap, H. Chua, M. J. K. Bashir et al., Central composite design (CCD) for parameters optimization of maximum power point tracking (MPPT) by response surface methodology (RSM), *Journal of mechanism of Continua and Mathematical Sciences*, 1, 1, 259–270, 2019.

[6] A. Ibrahim, S. Obukhov, and R. Aboelsaud, Determination of global maximum power point tracking of PV under partial shading using cuckoo search algorithm. *Applied Solar Energy*, 55, 367–375, 2020.

[7] C. Correa-Betanzo, H. Calleja, C. Aguilar et al., Photovoltaic-based DC microgrid with partial shading and fault tolerance. *Journal of Modern Power Systems and Clean Energy*, 7, 2, 340–349, 2019.

[8] M. Chen, S. Ma, J. Wu et al., Analysis of MPPT failure and development of an augmented nonlinear controller for MPPT of photovoltaic systems under partial shading conditions. *Applied Sciences*, 7, 1, 95–116, 2017.

[9] D. P. Mishra and S. Chakraborty, Application of soft computing in simulation of solar power tracking. In *Proceedings of 2018 Technologies for Smart-City Energy Security and Power (ICSESP), Bhubaneswar, India*, 2018, 1–5.

[10] L. Zhang, S. Yu, T. Fernando et al., An online maximum power point capturing technique for high-efficiency power generation of solar photovoltaic systems. *Journal of Modern Power Systems and Clean Energy*, 7, 2, 357–368, 2019.

[11] L. L. Jiang, R. Srivatsan, and D. L. Maskell, Computational intelligence techniques for maximum power point tracking in PV systems: A review. *Renewable and Sustainable Energy Reviews*, 85, 14–45, 2018.

[12] M. Séne, F. Ndiaye, M. E. Faye et al., A comparative study of maximum power point tracker approaches based on artificial neural network and fuzzy controllers. *International Journal of Physical Sciences*, 13, 1–7, 2018.

[13] M. G. Batarseh and M. E. Zater, Hybrid maximum power point tracking techniques: A comparative survey, suggested classification and uninvestigated combinations. *Solar Energy*, 169, 535–555, 2018.

4 Effective Efficiency Distribution Characteristics for Different Configurations of Arc and V-Shape Ribs in Solar Air Channels

A Comparative Study

Ashutosh Sharma and Ranchan Chauhan
Dr B R Ambedkar National Institute of Technology

CONTENTS

DOI: 10.1201/9781003272717-4

4.1 INTRODUCTION

The present scenario of the energy crisis and environmental degradation inculcates the requisite for research toward the effective utilization of renewable energy resources. Solar energy is the most prominent and abundantly available among the renewable energy resources. The directly convertible forms of solar energy that are being utilized with much higher intensities globally are thermal energy and electricity [1]. For a solar thermal energy process system, the solar collector is a vital constituent, which converts the solar radiations to the internal energy of a conveyance fluid. In most applications, air and water are used as conveyance mediums for the solar collector. However, the efficiency of solar air collectors is lower as compared to solar water collectors due to the poor thermophysical properties of air. Different methods have been suggested in the literature for improving the performance of solar air collectors, such as roughened absorber surface [2], jet impingement [3], and finned absorber plate [4]. Among these methods, the investigation of roughened absorber plate is widely studied due to the significant increase in the heat transfer rate with least possible friction losses [5]. Absorber plates roughened with ribs extend their application in many investigations of solar collectors and heat exchangers. The other methods of improving solar collector performance and effectiveness are tilt angle optimization [6] and utilization of thermal storage [7].

A deep literature survey suggests that various studies have been carried out for different arrangements of ribs, which enumerate their characteristics of heat transfer and flow conditions. Among these geometries, arc and V-shape ribs with different configurations have been studied mostly. Saini and Saini [8] investigated the performance of continuous arc shape ribs in the rectangular fluid flow channel. The effect of arc angle (α) and relative roughness height (e/D_h) was investigated for the range of Reynolds number of $2000 \leq \mathrm{Re} \leq 17000$. The improvement in Nu_{ca} and f_{ca} was found to be 3.8 and 1.75, respectively, from the conventional smooth plate air channel. The effect of discrete arc rib on heat transfer and fluid pattern in a solar air channel was studied by Hans et al. [9]. Experimentations were carried out in the Reynolds number range of $2000 \leq \mathrm{Re} \leq 16000$ by taking into account various geometrical parameters such as relative gap position (d/w), relative gap width (g/e), relative roughness pitch (P/e), e/D_h, and α. The maximum enhancement in Nu_{da} and f_{da} from the smooth plate was found to be 2.63 and 2.44, respectively. In another study, Singh et al. [10] investigated the performance of multiple arc ribs. The experimental study was carried out in the range of $2200 \leq \mathrm{Re} \leq 22000$. The augmentation in Nu_{ma} and f_{ma} was found to be 5.07 and 3.71 from the conventional solar air channel. As an extension to this, experimentations on discrete multiple arc ribs were conducted by Pandey et al. [11] for heat transfer and fluid flow pattern investigation. The experimentation was carried out in the Re range of 2100–21000. The extreme improvement in Nu_{dma} and f_{dma} was found to be 5.85 and 4.96, respectively. Various arrangements of V-shape rib geometries have also been studied in the literature. Momin et al. [12] investigated the performance of the continuous V-shape rib in the Reynolds number range of 2500–18000. The geometrical parameters considered were e/D_h and α. The enhancement in Nu_{cV} from the smooth plate was observed as 2.3. Another research was conducted by Singh et al. [13] to investigate the result of discrete V-shape ribs. Investigations were conducted in the Re range of 3000–15000 for parameters P/e, g/e, d/w, e/D_h, and α.

The augmentation in Nu_{dV} and f_{dV} compared to the smooth channel was 3.04 and 3.11, respectively. The thermo-hydraulic performance of the system was found to be significantly better than that of the flat plate solar air heater [14]. Singh et al. [15] also studied the effective efficiency characteristics of discrete V-shape ribs. Hans et al. [16] carried out experimentations to investigate the thermo-hydraulic performance of multiple V-shape ribs in the rectangular fluid flow channel. The experimentations were carried out in the Reynolds number range of 2000–20000 for parameters P/e, e/D_h, α, and relative roughness width (W/w). The authors reported an increment in Nu_{mV} and f_{mV} of 6 and 5 from Nu_{sc} and f_{sc}, respectively. In another investigation, the effect of discrete multiple V-shape ribs in the rectangular fluid flow channel was studied by Kumar et al. [17]. The investigation encompassed Re in the range of 2000–20000 for potential geometrical parameters of P/e, g/e, d/w, e/D_h, α, and W/w. The enhancement in Nu_{dmV} and f_{dmV} was observed as 6.74 and 6.37, respectively.

Apart from the arc and V-shape arrangements of ribs in the rectangular fluid flow channel, several other geometries have also been investigated. Wang et al. [18] studied the performance of "S"-shape ribs with the gap in the rectangular solar air channel. The main geometrical parameters of ribs considered for investigation was rib width, rib spacing, channel height, rib clearance width, air mass flow rate, and solar intensity. The authors reported that the performance increases by 48% from the conventional smooth plate solar air heater. The effect of the V-shape rib with systematic gap and the staggered rib was studied by Jain and Langewar [19]. Four absorber plates with varying P/e were investigated in a rectangular airflow channel, and the maximum improvement in friction factor and heat transfer was found to be 3.18 and 2.3, respectively, for $P/e = 12$. Kumar and Layek [20] carried out experimentations to find the effect of the twisted rib as roughness. The experimentations were conducted for the Re range of $3500 \leq Re \leq 21000$ considering geometrical parameters, rib inclination angle, relative roughness pitch, and twist ratio. The augmentation in Nu_s and f_s from the smooth plate was observed as 2.46 and 1.78, respectively. Ngo and Phu [21] investigated the performance of conic-curve profile ribs in the rectangular fluid flow channel. The authors found a significant increment in the Nusselt number and a decrement in the friction factor by decreasing the conic constant, which further results in the thermo-hydraulic performance parameter (THPP) improvement. The extreme value of THPP was found to be 1.9 at the conic constant of −4. Patel et al. [22] investigated the effect of reverse airfoil rib over the absorber plate on heat transfer and fluid flow characteristics. The authors studied the effect of the Reynolds number in the range of $6000 \leq Re \leq 18000$ on P/e and e/D_h. The authors reported the maximum value of THPP as 2.53 at Reynolds number 6000. Thakur and Thakur [23] experimentally analyzed the performance of W-shaped multi-staggered ribs in the solar air channel. The augmentation in the Nusselt number was observed to be between 2.3 and 4.1 from the smooth channel. In another study [24], the thermo-hydraulic performance parameter for the same roughness geometry was found to be 3.24. Jain et al. [25] experimentally assessed the performance of arc ribs with multiple gaps in the rectangular solar air channel. The experimentation was carried out for the Reynolds number range of 3000–18000. The augmentation in Nu and f was reported to be 274% and 169% from the smooth channel, respectively.

The brief literature review suggests that the arc and V-shape rib geometries have been of great interest for researchers and quite helpful for augmenting heat transfer with contemporary fewer friction losses. It also depicts that the geometrical potential parameters studied for different configurations of arc and V-shape ribs are almost similar. The distribution of the effective efficiency characteristics assesses the combined effect of heat transfer and fluid friction losses, which is much imperative for the overall efficiency estimation. So, in this study a comparative evaluation of effective efficiency characteristics for ribs of configurations continuous arc vs. V-shape, discrete arc vs. V-shape, multiple arc vs. V-shape, and discrete multiple arc vs. V-shape ribs has been carried out. For this, a mathematical model has been developed in MATLAB and validated. The effect of each parameter of the arc and V-shape rib configurations on effective efficiency characteristics has been studied. First, the similar configurations of arc and V-shape ribs have been compared individually, and afterward, all the selected configurations have been comparatively analyzed.

4.2 PERFORMANCE EVALUATION OF SOLAR AIR CHANNEL

The physics involved in the thermal conversion process of the solar air channel can be easily understood by the heat transfer phenomenon. In the solar air channel, the energy absorbed in the form of solar radiations is converted into the internal energy of the fluid, which resembles its functionality with a heat exchanger. For heat exchangers, the effective efficiency determination is critical to estimate both thermal and hydraulic performance. The schematic of the solar air collector is shown in Figure 4.1, with an effective heat transfer mechanism significant for efficiency assessment. To optimize the solar collector design by thermal and hydraulic component estimation, the prospective factors of researchers' interest are the useful heat gain, temperature profile variation, and convective heat transfer coefficient.

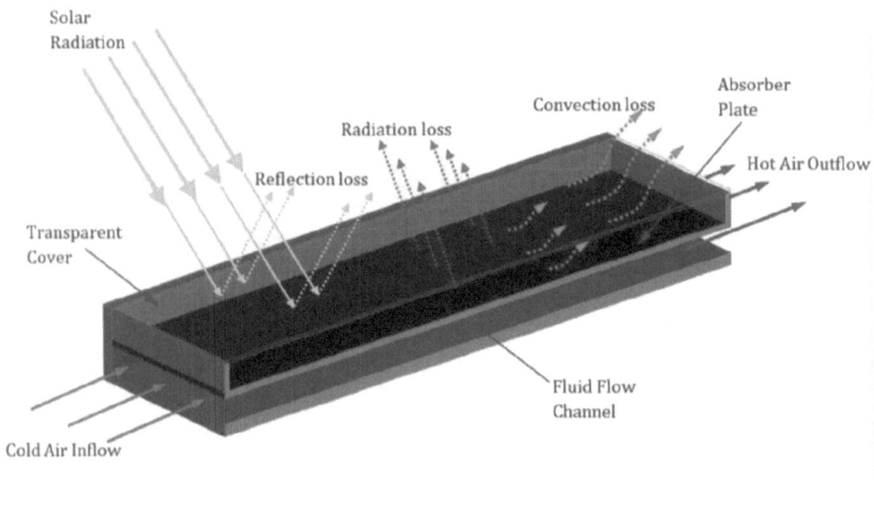

FIGURE 4.1 Side sectional isometric view of solar air collector with a heat transfer mechanism.

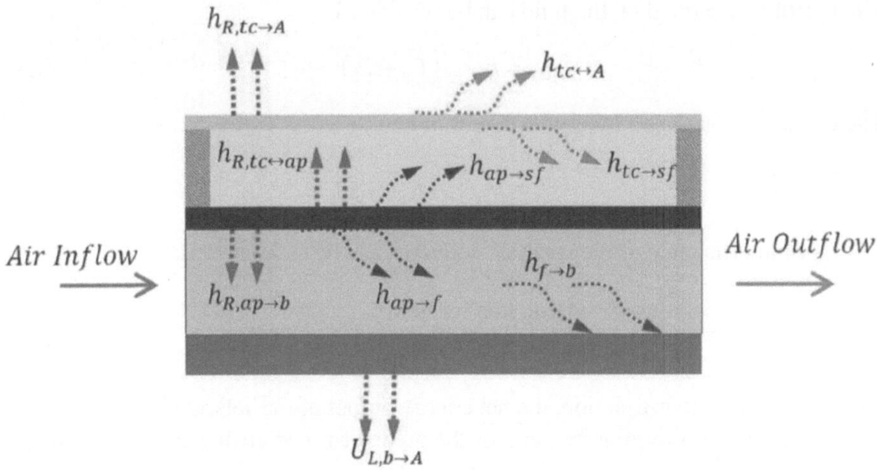

FIGURE 4.2 Energy balance in a solar air collector.

This section discusses the energy conservation equations necessary for effective efficiency attributes estimation.

The major assumption acknowledged for efficiency estimation is non-environmental interaction such as humidity and climatic variation with the solar collector. Figure 4.2 depicts that the incoming solar radiations fell over the transparent cover and passed to the absorber plate. Some amount of the energy is absorbed by the transparent cover, which is further convected and radiated to the stagnant fluid and surroundings. For this component, the energy balance equation is

$$h_{tc\leftrightarrow A}\left(T_{tc}-T_A\right)+h_{tc\to sf}\left(T_{tc}-T_{sf}\right)+h_{R,tc\leftrightarrow ap}\left(T_{tc}-T_{ap}\right)+h_{R,tc\to A}\left(T_{tc}-T_A\right)=I_t\alpha_{tc} \quad (4.1)$$

where

$$h_{R,tc\leftrightarrow ap}=\frac{\sigma\left(T_{tc}^2+T_{ap}^2\right)\left(T_{tc}+T_{ap}\right)}{\left(1/\varepsilon_{tc}+1/\varepsilon_{ap}-1\right)} \quad (4.2)$$

$$h_{R,tc\to A}=\frac{\sigma\left(T_{tc}^2+T_A^2\right)\left(T_{tc}+T_A\right)}{\left(1/\varepsilon_{tc}-1\right)} \quad (4.3)$$

The major portion of radiations received by the absorber plate is converted into the internal energy of the fluid passing beneath it, but simultaneously, some part of this energy is also convected and radiated to the stagnant fluid, transparent cover, and the base wall. The energy balance equation for the absorber plate is given as:

$$\begin{aligned}h_{ap\to sf}\left(T_{ap}-T_{sf}\right)+h_{ap\to f}\left(T_{ap}-T_f\right)+h_{R,tc\leftrightarrow ap}\left(T_{ap}-T_{tc}\right)+h_{R,ap\to B}\left(T_{ap}-T_B\right)\\=I_t\alpha_{ap}\tau_{tc}\end{aligned} \quad (4.4)$$

The useful heat gained by the fluid can be written as:

$$Q_u = h_{ap\rightarrow f}\left(T_{ap} - T_f\right) \tag{4.5}$$

The useful heat gain can otherwise be estimated by

$$Q_u = \dot{m}_a c_{p,a}\left(T_{ai} - T_{ao}\right) \tag{4.6}$$

The thermal efficiency of the solar air collector can be estimated as:

$$\eta_{th} = \frac{\text{Heat gain by fluid}}{\text{Radiations absorbed by absorber plate}} = \frac{Q_u}{I_t \times A_{f,ap}} \tag{4.7}$$

For effective energy estimation, the net energy output of the solar collector considering the pumping power requirement for the predetermined airflow is

$$E_{n,o} = Q_u - \frac{P_{\text{mech}}}{CF} \tag{4.8}$$

The pumping power is the function of pressure difference across the fluid flow channel and is given as:

$$P_{mech} = \frac{\dot{m}_a \times (\Delta p)_c}{\rho_a} \tag{4.9}$$

where

$$(\Delta p)_c = \frac{2f \times L_c \times G^2}{D_h \times \rho_a} \tag{4.10}$$

The effective efficiency of the solar air heater represents the ratio of net energy output to the incoming solar energy on the absorber surface. Mathematically,

$$\eta_{\text{effective}} = \frac{\text{Net energy output}}{\text{Radiations absorbed by absorber plate}} = \frac{Q_u - (P_{\text{mech}} / CF)}{I_t \times A_{f,ap}}$$
$$= \eta_{th} - \left[\frac{\rho_a f L_c V^3 (W_c + H_c)}{I_t \times A_{f,ap} \times CF}\right] \tag{4.11}$$

From equation 4.11, it is clear that the effective efficiency is always lesser in value than the thermal efficiency and considers the effect of both heat transfer and fluid friction in the channel.

4.3 EFFECTIVE EFFICIENCY EVALUATION

The effective efficiency of a solar collector considers both the thermal and hydraulic performance of a solar air heater. Therefore, the effect of each potential parameter of arc and V-shape configurations on the effective efficiency has been analyzed. For effective efficiency characteristics estimation of different geometries of arc rib in the rectangular solar air channel, a mathematical program has been formulated and validated corresponding to the respective experimental data. This section presents the mathematical model formulation procedure and validation methodology.

4.3.1 Mathematical Model

A mathematical model has been formulated in MATLAB for effective efficiency estimation. The basic assumptions made in this study corresponding to the solar collector are negligible flow conduction, negligible edge effects, and steady-state condition. The selection of potential parameters of different geometries of ribs has been done on the basis of respective studies reported in the literature. The mathematical program can be divided into four subparts, which are itemized further.

4.3.1.1 Program Initialization

The mathematical program is initialized by entering a value for constants and variables. The constants used in this study are L_c, H_c, W_c, K_i, ε_p, ε_g, V_w, I_t, T_{sun}, T_{ai}, N_{tc}, and D_h, whereas the variables are potential parameters of the respective rib geometry, such as e/D_h, P/e, α, W/w, g/e, and d/w. The T_{ai} is equal to T_A, and the T_{ao} is estimated based on temperature rise. The T_{am} is calculated from the average of T_{ai} and T_{ao}, and the respective air properties are calculated as [26]:

$$C_{p,a} = 1006\left(\frac{T_{am}}{293}\right)^{0.0155};\mu_a = 1.81\times10^{-5}\left(\frac{T_{am}}{293}\right)^{0.735};k_a = 0.0257\left(\frac{T_{am}}{293}\right)^{0.86} \quad (4.12)$$

The absorber plate mean temperature (T_{apm}) is initially set at 10°C more than T_{am}.

4.3.1.2 Useful Heat Gain Assessment

The useful heat gained by the thermal collector can be evaluated either by energy balance or by heat transfer estimation. In this code, first, the useful heat gain is evaluated by energy balance ($Q_{u,\,eb}$) and then by considering the effect of heat removal factor ($Q_{u,\,hr}$) and the final value is assumed when the percentage variation among these two becomes less than 0.1%. So as to determine the value of $Q_{u,\,eb}$, first, the total heat loss coefficient is calculated by

$$U_{Total} = U_T + U_B + U_E \quad (4.13)$$

The top, bottom, and edge thermal loss coefficients have been evaluated by [27]

$$U_T = \left[\frac{N_{tc}}{\frac{C_t}{T_{am}}\left(\frac{T_{am}-T_A}{N_{tc}+f_t}\right)^{0.33}} + \frac{1}{h_w}\right]^{-1} + \left[\frac{\sigma\left(T_{am}^2+T_A^2\right)\left(T_{am}+T_A\right)}{\frac{1}{\varepsilon_p+\left(0.05N_{tc}\left(1-\varepsilon_g\right)\right)}+\frac{2N_{tc}+f_t-1}{\varepsilon_g}-1}\right] \quad (4.14)$$

$$U_E = \frac{(W_c+L_c)\times L_i\times k_i}{W_c\times L_c\times t_e} \quad (4.15)$$

$$U_B = \frac{k_i}{t_i} \tag{4.16}$$

where

$$h_w = 5.7 + 3.8V_w; C_t = 365.9\left[1 - 0.00883\beta + 0.0001298\beta^2\right]; f_t = \left(1 - 0.04h_w + 0.0005h_w^2\right)\left(1 + 0.091N_{tc}\right) \tag{4.17}$$

Finally, the $Q_{u,\,eb}$ is calculated by

$$Q_{u,eb} = \left[\left(I_t\alpha_{ap}\tau_{tc}\right) - U_{\text{total}}\left(T_{apm} - T_A\right)\right] \times A_{f,ap} \tag{4.18}$$

From $Q_{u,\,eb}$, the values of Re and $\dot{m}_a$ are estimated:

$$\text{Re} = \frac{GD_h}{\mu_a}; \dot{m}_a = \frac{Q_{u,eb}}{C_{p,a} \times \Delta T} \tag{4.19}$$

Afterward, the value of the convective heat transfer coefficient (h) is computed from the correlations for Nu as shown in Table 4.1. The schematics of the different configurations of arc and V-shape ribs with notations are shown in Figures 4.3 and 4.4, respectively.

After calculating h, the plate efficiency factor ($F_{p\eta}$) and heat removal factor (F_{hr}) have been evaluated by the following correlations:

$$F_{p\eta} = \frac{h}{h + U_{\text{Total}}}; F_{hr} = \frac{\dot{m}C_{p,a}}{A_{f,ap}U_{\text{Total}}}\left[1 - \exp\left(-\frac{A_{f,ap}U_{\text{Total}}F_{p\eta}}{\dot{m}C_{p,a}}\right)\right] \tag{4.20}$$

The useful heat gain considering the heat removed from a solar collector ($Q_{u,\,hr}$) is calculated by

$$Q_{u,hr} = F_{hr} \times A_{f,ap} \times \left[\left(I_t\tau_{tc}\alpha_{ap}\right) - U_{Total}\left(T_{apm} - T_A\right)\right] \tag{4.21}$$

The value of $Q_{u,\,hr}$ is compared with $Q_{u,\,eb}$, and if the percentage variation is found to be more than 0.1%, then T_{apm} is computed from $Q_{u,\,hr}$ using correlation:

$$T_{apm} = T_A + \left[\frac{\left(I_t\tau_{tc}\alpha_{ap}\right) - \left(Q_{u,hr} / A_{f,ap}\right)}{U_{\text{Total}}}\right] \tag{4.22}$$

From this T_{apm}, the $Q_{u,\,eb}$ is estimated until the convergence criteria are satisfied. The final value of $Q_{u,\,hr}$ ($\approx Q_{u,\,eb}$) is used for effective efficiency determination.

4.3.1.3 Effective Efficiency Assessment

The effective efficiency considers the effect of both the useful heat gain by the solar collector and the mechanical power required. For mechanical power (P_{mech}) assessment, equation 4.9 has been utilized and $(\Delta p)_c$ has been derived from equation 4.10. The values of $\dot{m}_a$, Re, and G corresponding to the final iteration are used in

TABLE 4.1
Correlation for Different Geometries of Ribs

S. No.	Geometry	Correlations
1	Arc ribs	$Nu_{ca} = 0.001047\,\mathrm{Re}^{1.3186}(e/d)^{0.3772}(\alpha/90)^{-0.1198}$ $f_{ca} = 0.14408\,\mathrm{Re}^{-0.17103}(e/d)^{0.1765}(\alpha/90)^{0.1185}$
2	Discrete arc ribs	$Nu_{da} = 1.014\times10^{-3}\,\mathrm{Re}^{1.036}(P/e)^{2.522}(\alpha/90)^{-0.293}(d/w)^{-0.078}\times(g/e)^{-0.016}\times(e/D_h)^{0.412}\times\exp\left(-0.567(\ln(P/e))^2\right)$ $\times\exp\left(-0.114(\ln(\alpha/90))^2\right)\times\exp\left(-0.077(\ln(d/w))^2\right)\exp\left(-0.133(\ln(g/e))^{220527}\right)$ $f_{da} = 8.1921\times10^{-2}(\mathrm{Re})^{-0.147}(P/e)^{1.191}(\alpha/90)^{-0.292}(d/w)^{-0.067}\times(g/e)^{-0.006}(e/D_h)^{0.528}$ $\exp\left(-0.255(\ln(P/e))^2\right)\exp\left(-0.110(\ln(\alpha/90))^2\right)\times\exp\left(-0.063(\ln(d/w))^2\right)\exp\left(-0.158(g/e)^2\right)$
3	Multiple arc ribs	$Nu_{ma} = 1.564\times10^{-4}\,\mathrm{Re}^{1.343}(e/D)^{0.048}(W/w)^{0.407}\times\exp\left(-0.099(\ln(W/w))^2\right)(P/e)^{0.572}$ $\times\exp\left(-0.148(\ln(P/e))^2\right)\times(\alpha/90)^{-0.355}\times\exp\left(-0.272(\ln((\alpha/90)))^2\right)$ $f_{ma} = 0.063\times(e/D)^{0.102}\times\mathrm{Re}^{-0.16}\times(W/w)^{0.277}\times(p/e)^{0.562}\times(\alpha/90)^{0.023}\times\exp\left(-0.140(\ln(p/e))^2\right)\times\exp\left(-0.013(\ln(\alpha/90))^2\right)$
4	Discrete multiple arc ribs	$Nu_{dma} = 1.39\times10^{-4}\times\mathrm{Re}^{1.3701}\times(d/x)^{-0.4997}\times(g/e)^{-0.0292}\times(W/w)^{0.4017}\times(p/e)^{0.5854}\times(\alpha/60)^{-2.235}\times(e/D)^{0.0931}$ $\times\exp\left(-0.3989(Ln(d/x))^2\right)\times\exp\left(-0.2013(Ln(g/e))^2\right)\times\exp\left(-0.129(\ln(W/w))^2\right)\times\exp\left(-0.142(\ln(P/e))^2\right)\times\exp\left(-0.5614(\ln(\alpha/60))^2\right)$ $f_{dma} = 2.11\times10^{-1}\times\mathrm{Re}^{-0.25}\times(d/x)^{-0.888}\times(g/e)^{-0.079}\times(W/w)^{0.032}\times(p/e)^{0.643}\times(\alpha/60)^{-2.546}\times(e/D)^{0.145}$ $\times\exp\left(-0.662(Ln(d/x))^2\right)\times\exp\left(-0.496(\ln(g/e))^2\right)\times\exp\left(-0.160(\ln(p/e))^2\right)\times\exp\left(-3.96(\ln(\alpha/90))^2\right)$

(Continued)

TABLE 4.1 (*Continued*)
Correlation for Different Geometries of Ribs

S. No.	Geometry	Correlations
5	V-shape ribs	$Nu_{cV} = 0.067 \times \text{Re}^{0.888} (e/D_h)^{0.424} (\alpha/60)^{-0.077} \times \exp\left[-0.782 \times (\ln \alpha/60)^2\right]$
		$f_{cV} = 6.266 \times \text{Re}^{-0.425} \times (e/D_h)^{0.565} \times (\alpha/60)^{-0.093} \times \exp\left(-0.719(\ln(\alpha/60))^2\right)$
6	Discrete V-shape ribs	$Nu_{dV} = 2.36 \times 10^{-3}\, \text{Re}^{0.9} (P/e)^{3.5} (\alpha/60)^{-0.23} (d/w)^{-0.043} \times (g/e)^{-0.014} \times (e/D_h)^{0.47} \times \exp\left(-0.84(Ln(P/e))^2\right)$ $\times \exp\left(-0.72(\ln(\alpha/60))^2\right) \times \exp\left(-0.05(\ln(d/w))^2\right) \exp\left(-0.15(\ln(g/e))^{220527}\right)$
		$f_{dV} = 4.13 \times 10^{-2}\, \text{Re}^{-0.126} \times (P/e)^{2.74} \times (\alpha/60)^{-0.034} \times (d/w)^{-0.058} \times (g/e)^{0.031} \times (e/D_h)^{0.70} \times \exp\left(-0.685(\ln(P/e))^2\right) \times \exp\left(-0.93(\ln(\alpha/60))^2\right)$ $\times \exp\left(-0.058(\ln(d/w))^2\right) \exp\left(-0.21(\ln(g/e))^{220527}\right)$
7	Multiple V-shape ribs	$Nu_{mV} = 3.35 \times 10^{-5}\, \text{Re}^{0.92} (e/D)^{0.77} (W/w)^{0.43} (\alpha/90)^{-0.49} \times \exp^{\left(-0.117(\ln(W/w))^2\right)} \times \exp^{\left(-0.61(\ln(\alpha/90))^2\right)} (P/e)^{8.54} \times \exp\left(-2.0407(\ln(P/e))^2\right)$
		$f_{mV} = 4.47 \times 10^{-4}\, \text{Re}^{-0.3188} \times (e/D)^{0.73} \times (W/w)^{0.22} \times (\alpha/90)^{-0.39} \times \exp^{\left(-0.52(\ln(\alpha/90))^2\right)} \times (P/e)^{8.9} \exp^{\left(-2.133(\ln(P/e))^2\right)}$
8	Discrete multiple V-shape ribs	$Nu_{dmV} = 8.532 \times 10^{-3} \times \text{Re}^{0.932} \times (e/D)^{0.175} \times (W/w)^{0.506} \times \exp\left(-0.0753(\ln(W/w))^2\right)(G_d/L_v)^{-0.0348} \times \exp\left(-0.0653(\ln(G_d/L_v))^2\right)(g/e)^{-0.0708}$ $\times \exp\left(-0.223(\ln(g/e))^2\right)(\alpha/60)^{-0.0239} \times \exp\left(0.1153(\ln(\alpha/60))^2\right)(P/e)^{1.196} \times \exp\left(-0.2805(\ln(p/e))^2\right)$
		$f_{dmV} = 3.19 \times \text{Re}^{-0.3151} \times (e/D)^{0.268} \times (W/w)^{0.1132} \times \exp\left(-0.0974(\ln(W/w))^2\right)(G_d/L_v)^{0.0610} \times \exp\left(-0.1065(\ln(G_d/L_v))^2\right)(g/e)^{-0.1769}$ $\times \exp\left(-0.6349(\ln(g/e))^2\right)(\alpha/60)^{0.1553} \times \exp\left(-0.1527(\ln(\alpha/60))^2\right)(P/e)^{-0.7941} \times \exp\left(0.1486(\ln(p/e))^2\right)$

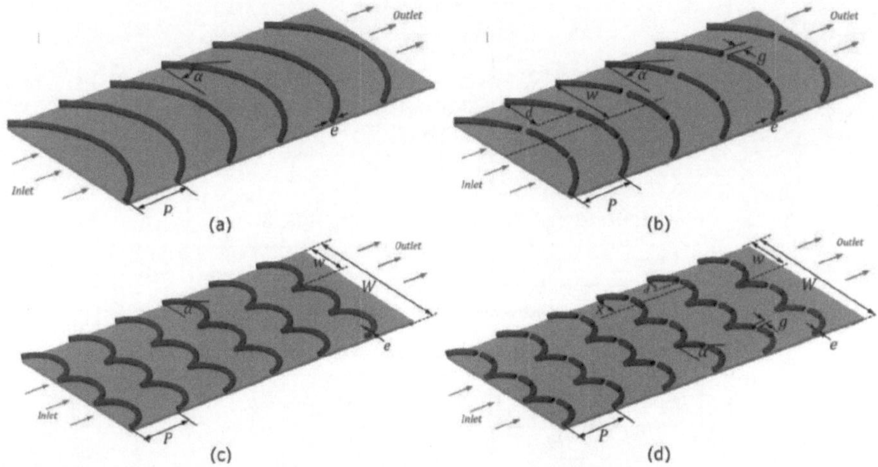

FIGURE 4.3 Schematic of (a) continuous arc, (b) discrete arc, (c) multiple arc, and (d) discrete multiple arc ribs with notations.

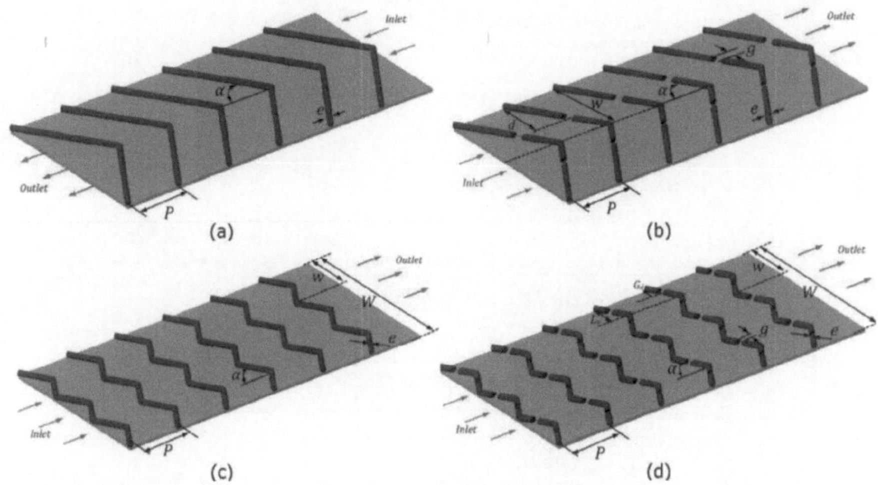

FIGURE 4.4 Schematic representing (a) continuous V-shape, (b) discrete V-shape, (c) multiple V-shape, and (d) discrete multiple V-shape ribs with notations.

correlations. The value of f is estimated from correlations provided in the literature given in Table 4.1. After estimating the P_{mech}, the thermal and mechanical efficiencies have been estimated using equations 4.7 and 4.11, respectively. The flow chart of the MATLAB program is shown in Figure 4.5 for a better understanding of the effective efficiency estimation methodology. The ranges of potential parameters that have been used as variables in this model are presented in Table 4.2.

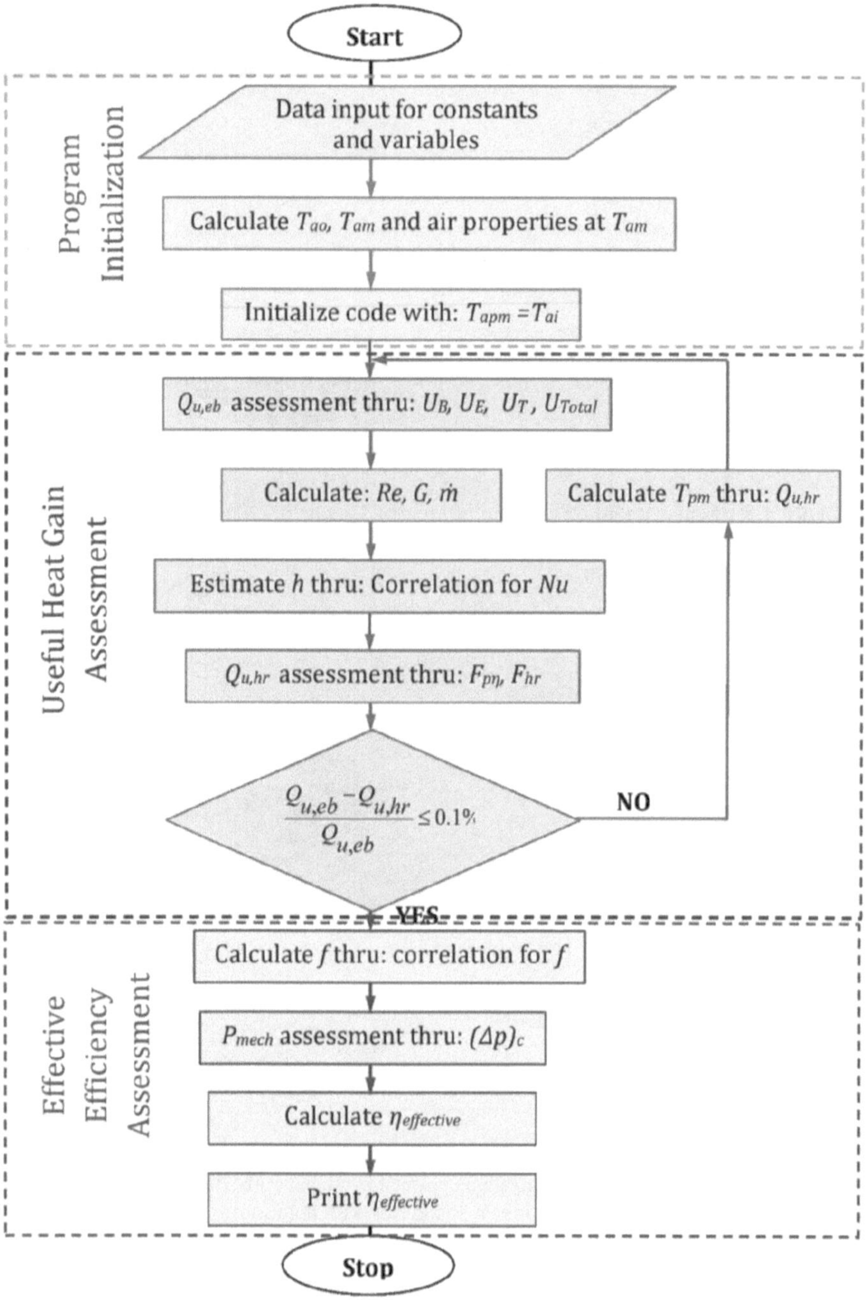

FIGURE 4.5 Flow chart for the computational program of effective efficiency characteristics estimation.

TABLE 4.2
Range of Potential Parameters for Different Arrangements of Ribs

S. No.	Arrangements of Ribs	e/D_h	α (Degree)	P/e	d/w	W/w	g/e	Aspect Ratio (W/H)
1	Continuous arc ribs	0.02; 0.022; 0.027; 0.035	30; 45; 60	10	-	-	-	12
2	Discrete arc ribs	0.022; 0.027; 0.035; 0.039; 0.043	30; 45; 60; 75	4; 6; 8; 10; 12	0.2; 0.35; 0.5; 0.65; 0.8	-	0.5; 1.0; 1.5; 2.0; 2.5	12
3	Multiple arc ribs	0.018; 0.027; 0.035; 0.045	30; 45; 60; 75	6; 8; 10; 12	-	1; 3; 5; 7	-	11
4	Discrete multiple arc ribs	0.018; 0.027; 0.035; 0.045	30; 45; 60; 75	4; 8; 12; 16	0.25; 0.45; 0.65; 0.85	1; 3; 5; 7	0.5; 1.0; 1.5; 2.0; 2.5	10
5	Continuous V-shape ribs	0.02; 0.022; 0.027; 0.035	30; 45; 60	10	-	-	-	10.15
6	Discrete V-shape ribs	0.022; 0.027; 0.035; 0.039; 0.043	30; 45; 60; 75	4; 6; 8; 10; 12	0.2; 0.35; 0.5; 0.65; 0.8	-	0.5; 1.0; 1.5; 2.0; 2.5	12
7	Multiple V-shape ribs	0.018; 0.027; 0.035; 0.045	30; 45; 60; 75	6; 8; 10; 12	-	2; 4; 6; 8	-	12
8	Discrete multiple V-shape ribs	0.018; 0.027; 0.035; 0.045	30; 45; 60; 75	4; 8; 12; 16	0.25; 0.45; 0.65; 0.85	2; 4; 6; 8	0.5; 1.0; 1.5; 2.0; 2.5	10

4.3.2 Mathematical Model Validation

The validation of the developed simulation code has been carried out by comparing the values of Nusselt number and friction factor obtained from the code with experimental results of respective studies. The values of potential parameters used for validation are shown in Table 4.3. The simulation code with these fixed values of potential parameters for each geometry has been run, and the obtained results for Nusselt number and friction factor are presented in comparison with the experimental results in Figures 4.6 and 4.7. The maximum percentage variations among the obtained results and experimental results are found to be $\pm 3.64\%$ and $\pm 4.29\%$ for Nusselt number and friction factor, respectively. The validation clarifies that the results obtained from the simulation code are in good agreement with the experimental data with the acceptable percentage variation ($\leq 5\%$).

4.4 RESULTS AND DISCUSSION

The effective efficiency characteristics of different rib geometries including continuous arc and V-shape, discrete arc and V-shape, multiple arc and V-shape, and discrete multiple arc and V-shape ribs have been estimated from the mathematical simulation model discussed in the last section. The effects of potential parameters have been investigated, and the results have been discussed in this section. For proportional performance evaluation of these geometries, a comparative analysis has also been carried out for selected configurations with optimum potential parameters suggested by the literature study.

4.4.1 Continuous Arc and V-Shape Ribs

The performance of the continuous arc and V-shape ribs have been comparatively investigated by evaluating the effect of their geometrical potential parameters on effective efficiency characteristics. Figure 4.8 shows that as the Reynolds number increases, the slope of the effective efficiency curve grows positive, but

TABLE 4.3
Values of Potential Parameters Used for Validation

S. No.	Potential Parameter	Value
1	Re	2000–21000
2	e/D_h	0.043
3	A	60
4	P/e	10
5	d/w	0.65
6	W/w	5
7	g/e	1.0

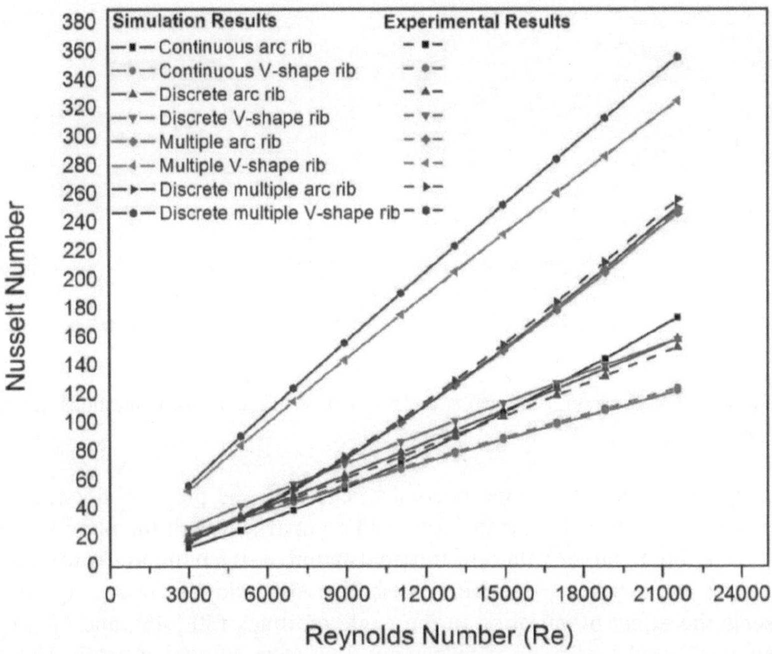

FIGURE 4.6 Comparison of predicted values of Nusselt number from the simulation code and experiments.

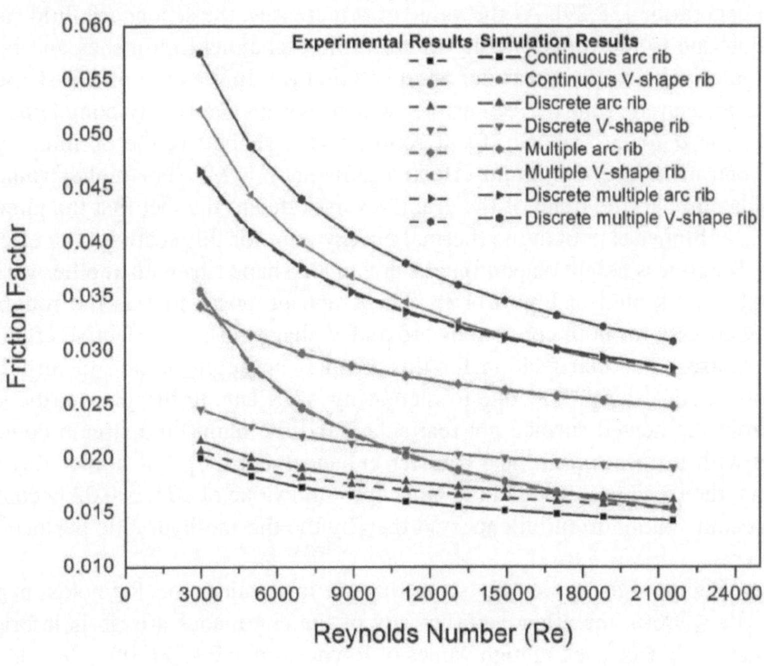

FIGURE 4.7 Comparison of friction factor obtained from simulation code and experimentation.

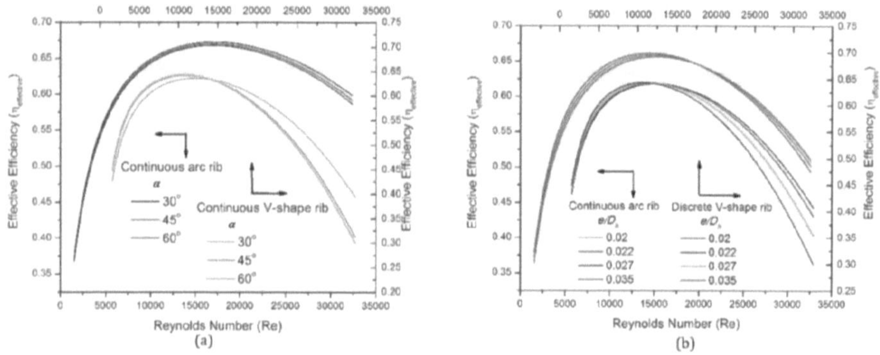

FIGURE 4.8 Effect of (a) α and (b) e/D_h on effective efficiency of continuous arc and V-shape ribs.

after a specific value, the slope becomes negative. The positive slope is due to the enhancement in the heat transfer rate in comparison with the pumping power requirement, but at larger values of Reynolds number, the pumping power requirement increases; therefore, the overall effective efficiency decreases. Figure 4.8a represents the effect of variation in the angle of attack (30°, 45°, and 60°) on the continuous arc and V-shape ribs for the constant value of e/D_h at 0.043. The graph shows that the effective efficiency decreases with its increasing value. This is due to the fact that at lesser values of α for arc rib, its effect can be similarized as transverse rib, which has a high heat transfer rate because of secondary vortex fluid stream generation [28,29]. As the value of α increases, the secondary fluid stream formation due to the transversality of rib in fluid channel diminishes and results in comparatively less heat transfer augmentation [2]. In the case of a V-shape rib, the α is measured from the centerline, which is a geometrically complementary angle to the α for arc rib. So, for a continuous V-shape rib, the optimum value corresponding to the maximum effective efficiency is 60°. For higher values of Reynolds number, the trend of the graph reverses due to the fact that the pumping power is at higher play than the thermal energy gain for this section. The effect of relative roughness height on continuous arc and V-shape ribs with another variable constant is presented in Figure 4.8b. The graph proposes that as the roughness height increases for both continuous arc and V-shape ribs, the effective efficiency also increases. The clarification for this trend is an augmentation in turbulence with more protrusion inflow due to increasing e/D_h. Due to turbulence, the shear layer from the heated surface got reattached and the main fluid stream comes in contact with it, resulting in heat transfer enhancement [30]. For higher Reynolds numbers, the trend reverses and the most optimum value of e/D_h is 0.02 because of its least contribution to turbulence and thereby the thermo-hydraulic performance enhances.

From Figure 4.8, it can be seen that at low values of Reynolds number ($3000 \leq Re \leq 7000$), the effective efficiency of the continuous arc rib is inferior to the V-shape rib. However, at high values of Reynolds number ($21000 \leq Re \leq 32000$), the slope of effective efficiency for continuous arc ribs is lesser than that for V-shape ribs. Therefore, the effective efficiency for continuous arc ribs at this Reynolds number range is higher than for the continuous V-shape ribs.

4.4.2 Discrete Arc and V-Shape Ribs

For discrete arc and V-shape ribs, the effective efficiency characteristics for variation in potential parameters α, e/D_h, d/w, g/e, and P/e have been assessed. Figure 4.9a represents the effect of variation in α (30°, 45°, 60°, 75°) on effective efficiency w.r.t. Reynolds number and other variable constants. The most optimum value of α corresponding to arc and V-shape ribs is 30° and 60°, respectively, for low Reynolds number range, and at higher range, the trend reverses. The explanation for the attainment of these optimum values can be referred to as continuous arc and V-shape ribs. The effect of variation in d/w on $\eta_{effective}$ for discrete arc and V-shape ribs with another variable constant is plotted in Figure 4.9b. The plot depicts the increment in the effective efficiency for both discrete arc and V-shape ribs as the d/w increases and attains the maximum value at 0.65 for low Reynolds number range. Beyond this point, the effective efficiency starts decreasing. The presence of ribs creates secondary rotating fluid streams in the flow region, and through the discrete slice of rib, this fluid stream escapes and mixes with the mainstream of fluid. The mixing of these two streams causes turbulence and results in heat transfer augmentation. When the discrete slice is too near the apex, it may not contribute much energy to fluid mixing and therefore the heat transfer augmentation is low in that case. The effective efficiency increases as the position of the discrete slice approaches the centerline until d/w becomes 0.65. Beyond this point, due to the increased length of rib resulting in the truncation of secondary fluid contribution, fluid mixing becomes insignificant again and the heat transfer enhancement starts decreasing again. For a higher range of Reynolds number, the trend reverses due to the overwhelming effect of pumping power than the heat transfer augmentation. The effect of variation in e/D_h on effective efficiency at different Reynolds numbers with another variable constant is plotted in Figure 4.9c. It can be evident that for both discrete arc and V-shape ribs, the effective efficiency increases as e/D_h increases and attains the maximum value at 0.043 for low Reynolds number range. The maximum effective efficiency corresponding to a maximum value of e/D_h is in good agreement with the results as obtained for continuous arc and V-shape ribs. Figure 4.9d depicts the effect of variation in relative gap width (g/e) on $\eta_{effective}$ at different Reynolds numbers for discrete arc and V-shape ribs with other variables kept constant. The g/e is varied through five values: 0.5, 1.0, 1.5, 2.0, and 2.5, and the corresponding effect on effective efficiency has been noted. The plot shows that as the relative gap width increases, the $\eta_{effective}$ also increases till its value becomes 1.0, but beyond this point, the effective efficiency again starts decreasing for the low Reynolds number range. As the secondary fluid stream passes through the discrete slice, two factors having a contrasting effect on each other and governing the energy of this stream are the area of discrete slice and the tangential velocity of the secondary stream. For a larger discrete slice area, the fluid velocity decreases and vice versa. So, due to these two factors, an average value of discrete slice provides a maximum effective efficiency. At a higher Reynolds number, the pumping power effect becomes higher than that of the heat transfer augmentation and therefore the

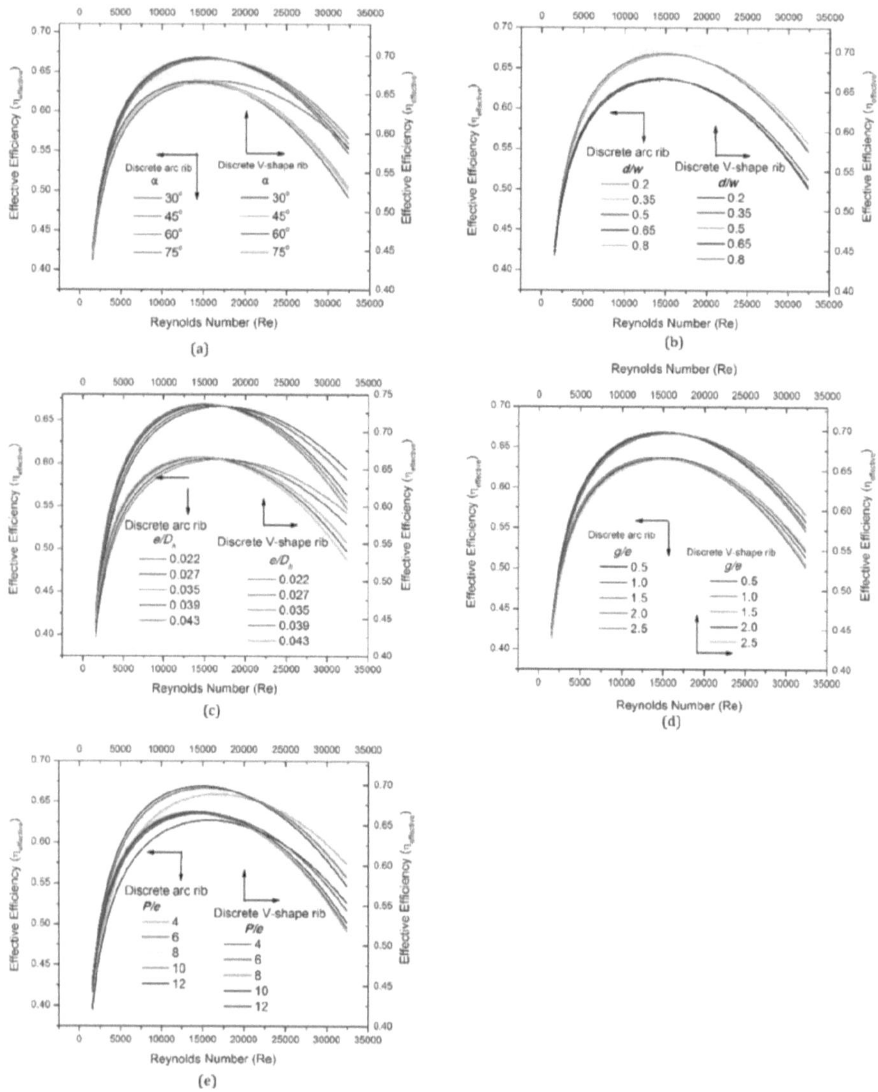

FIGURE 4.9 Effect of (a) α, (b) d/w, (c) e/D_h, (d) g/e, and (e) P/e on effective efficiency for discrete arc and V-shape ribs.

trend reverses. The variation effect of relative roughness pitch (P/e) and other constant variables on effective efficiency corresponding to discrete arc and V-shape ribs is presented in Figure 4.9e. For both discrete arc and V-shape ribs, the effective efficiency first increases with an increase in P/e and, after reaching a certain value, it starts decreasing for the low Reynolds number range. As the fluid flows past a rib, the flow gets deviated toward the absorber surface and the flow reattachment point is created. At this point, the heat transfer rate is maximum, and beyond this point toward downstream rib, the boundary layer develops, resulting in a lower convective heat transfer. So, the increase in the pitch heat transfer rate also reduces and results in a lower effective efficiency. For small values of pitch, the flow reattachment does not occur, so the effective efficiency is again low. The optimum value of P/e for discrete arc and V-shape ribs is 10 and 8, respectively, for a low Reynolds number range, and for a higher range, the trend reverses. The difference in these two optimum values is significant for practical implementation because of material and manufacturing cost association. So, an in-depth comparative analysis is required to be performed for finding the cause of such deviations.

The effective efficiency for discrete arc and V-shape ribs also shows the same trend as that of continuous arc and V-shape ribs. The effective efficiency of discrete arc ribs has been found inferior to the discrete V-shape ribs at the low range of Reynolds number ($3000 \leq Re \leq 7000$). The same trend is also continuous at the medium range of Reynolds number ($7000 \leq Re \leq 25000$). However, at a high Reynolds number ($25000 \leq Re \leq 32000$), a higher effective efficiency was observed for discrete arc ribs than for discrete V-shape ribs. These results necessitate the comparative in-depth CFD analysis of these roughness geometries.

4.4.3 Multiple Arc and V-Shape Ribs

The effective efficiency characteristics of multiple arc and V-shape ribs have been investigated for four potential parameters, viz. α, e/D_h, P/e, and W/w. The variational effect of α on $\eta_{effective}$ w.r.t. Re for multiple arc and V-shape ribs with constant values of other variables is presented in Figure 4.10a. For multiple arc and V-shape ribs, the optimum value of α is found to be 30° and 60°, respectively, for low Reynolds numbers and at higher values of Re, the trend overturns. Figure 4.10b depicts the effect of e/D_h on $\eta_{effective}$ and its maximum value is observed corresponding to 0.045 with other variables kept constant. These results of α and e/D_h for multiple arc and V-shape ribs are found to be in good agreement with the results of previously studied rib configurations. In another case, the effect of P/e on effective efficiency is presented in Figure 4.10(c). The most optimum values both for multiple arc and V-shape ribs are found to be 8 for low Reynolds number, and at higher values of Re, the trend overturns. The effect of relative roughness width (W/w) on effective efficiency at variable Reynolds number for multiple arc and V-shape ribs is shown in Figure 4.10d with other variables constant. It can be evident that the effective efficiency increases with an increase in W/w up to a certain value and starts decreasing afterward. By increasing the arcs and V's in the channel, the secondary flow formation on the upstream side of the rib increases and therefore the heat transfer rate also increases, resulting in the effective efficiency augmentation. By increasing the arcs and V's beyond a certain limit, secondary flow streams increase, but their energy level decreases due to the flow over a very small rib region; therefore, the heat transfer rate decreases and friction factor increases. The most optimum value of W/w for multiple arc and V-shape ribs are found to be 6 and 5, respectively, for low Reynolds numbers, and at a higher range, the trend overturns due to the higher play of pumping power than thermal gain in this region. The difference in the optimum values of two geometries signifies the requirement of an in-depth analysis for understanding the physics more accurately. The effective efficiency can be higher for multiple arc ribs than for V-shape ribs at low

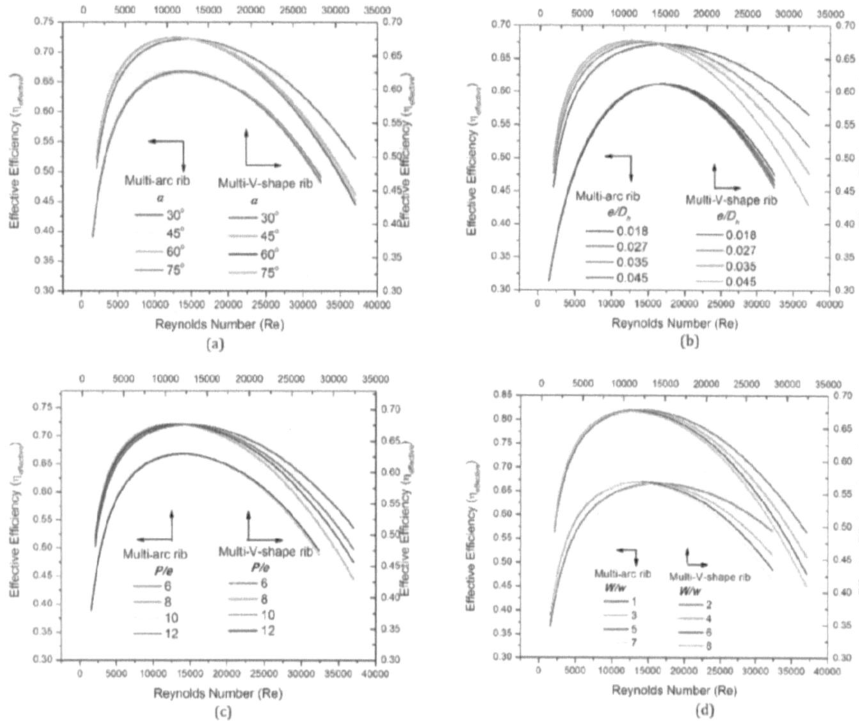

FIGURE 4.10 Effect of (a) α, (b) e/D_h, (c) P/e, and (d) W/w on effective efficiency of multiple arc and V-shape ribs.

and medium range of Reynolds numbers ($3000 \leq Re \leq 25000$). At a higher Reynolds number, this trend reverses and high efficiency is observed for multiple arc ribs.

4.4.4 Discrete Multiple Arc and V-Shape Ribs

The effective efficiency characteristics of the discrete multiple arc and V-shape ribs for potential parameters, viz. α, d/w, e/D_h, g/e, P/e, and W/w, with variation in the Reynolds number have been investigated. Figure 4.11a depicts the effect of variation in α on $\eta_{effective}$ w.r.t. Reynolds number at a constant value of other variables. The optimum value for discrete multiple arc and V-shape ribs is found to be 30° and 60°, respectively, for low Reynolds number, and at higher values, the trend reverses. Similarly, the effect of d/w on the effective efficiency of these geometries at the constant value of other variables is presented in Figure 4.11b. For lower values of Reynolds number, the maximum effective efficiency is observed at 0.65 both for discrete multiple arc and V-shape ribs, and at the higher range, the trend reverses. Similar trends of graphs for e/D_h, g/e, P/e, and W/w are found for discrete multiple arc and V-shape ribs as that of previously studied geometries with the same optimum values as shown in Figure 4.11c–f, respectively. The effective efficiency for all parameters can be more for discrete multiple V-shape ribs than for arc ribs at low

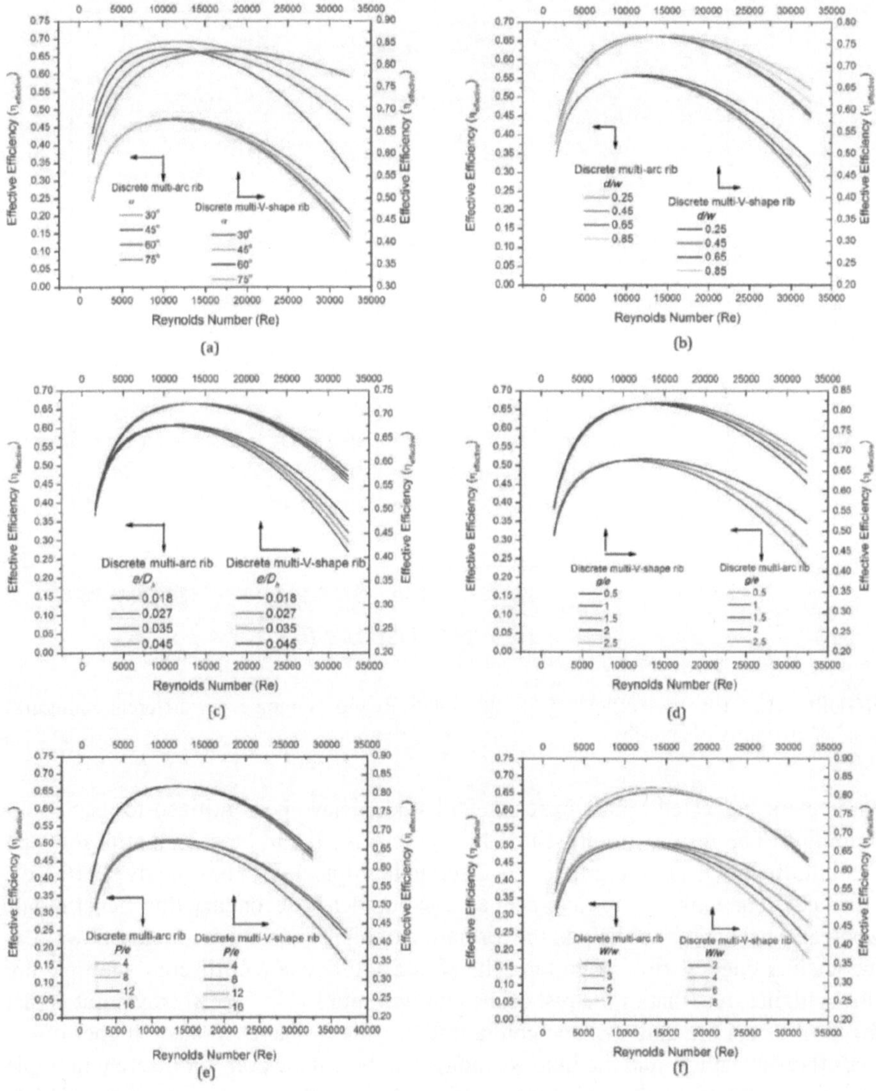

FIGURE 4.11 Effect of (a) α, (b) d/w, (c) e/D_h, (d) g/e, (e) P/e, and (f) W/w on $\eta_{\text{effective}}$ for discrete multiple arc vs V-shape ribs.

Reynolds numbers. At a higher Reynolds number range, this trend reverses and the effective efficiency of discrete multiple arc ribs increases compared to discrete multiple V-shape ribs.

4.4.5 Intercomparison

The performance of the selected arc and V-shape rib configurations has been compared with each other at different Reynolds numbers. The optimum potential

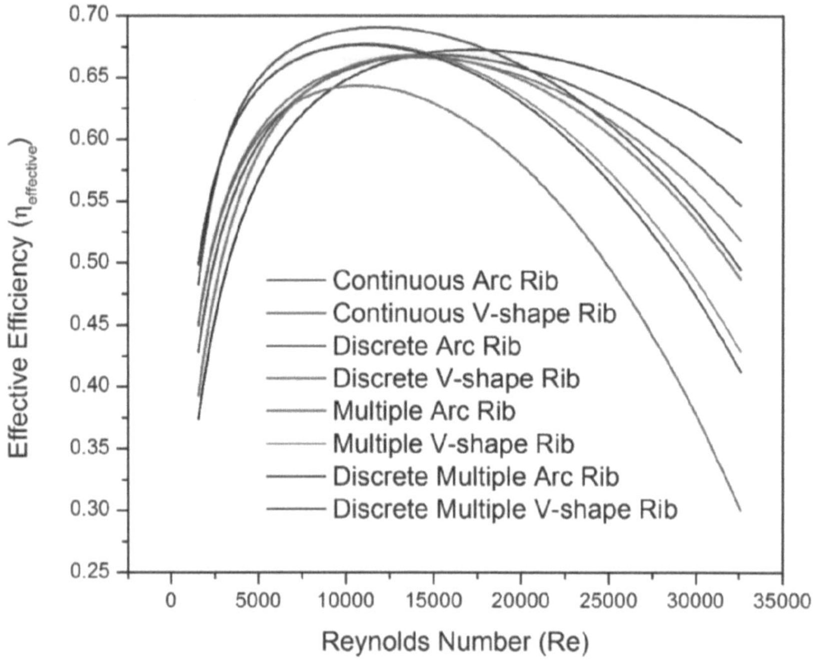

FIGURE 4.12 Effective efficiency variation with Reynolds number for different configurations of arc and V-shape ribs.

parameters suggested by the experimental studies have been utilized for each configuration. The results are presented in Figure 4.12. It can be seen that at low and medium Reynolds number range, the discrete multiple arc rib has the most effective efficiency. The multiple ribs augment a large number of secondary flow generations, whereas the discrete ribs help in the formation of high-intensity secondary flows near the leading edge of ribs. Therefore, the effective efficiency of discrete multiple arc ribs is higher than that of the rest of the configurations. The noteworthy point is that for most of the arc and V-shape configurations, the V-shape rib has a higher effective efficiency at low and medium Reynolds numbers. However, for discrete multiple ribs, this trend reverses and a higher efficiency was observed for discrete multiple arc ribs than for V-shape ribs. A more in-depth analysis is required for finding the cause of this trend. For the large Reynolds number range, the effective efficiency trend changes and its maximum value is observed for continuous arc ribs. This trend is due to the least contribution of continuous arc shape ribs to turbulence promotion at a higher Reynolds number.

4.5 CONCLUSIONS

The comparative investigation for various arc and V-shape rib configurations, viz. continuous arc vs continuous V-shape, discrete arc vs discrete V-shape, multiple arc vs multiple V-shape, and discrete multiple arc vs discrete multiple V-shape, has been carried out. The effective efficiency characteristics have been studied for comparatively evaluating the effect of each geometrical parameter. The following broad conclusions have been drawn:

1. The effective efficiency increases with an increase in the Reynolds number for a certain limit (*15000±3000*) and afterward starts decreasing because of the diminishing effect of thermal energy gain as compared to an abrupt increase in pumping power requirement.
2. The optimum geometrical parameter sets suggested by experimental results from previous studies are only valid within a certain range of Reynolds number, especially for the lower Reynolds number range ($3000 \leq Re \leq 12000$). For the higher Reynolds number ($15000 \leq Re \leq 30000$) range, the optimum value of potential parameter changes because of the dominating effect of pumping power requirement over thermal energy gain.
3. The effective efficiency for selected arc and V-shape configurations is higher for V-shape ribs as compared to arc shape ribs at the low ($3000 \leq Re \leq 7000$) and medium Reynolds number range ($7000 \leq Re \leq 25000$). For the higher Reynolds number range ($25000 \leq Re \leq 32000$), the lower effective efficiency is observed for V-shape ribs as compared to arc rib configurations.
4. For selected configurations of arc and V-shape ribs, the maximum effective efficiency in the low and medium Reynolds number range is observed at $e/D_h = 0.045$, $g/e = 1$, and $d/w = 0.65$. The optimum values of α for configurations of arc and V-shape ribs are found to be 30° and 60°, respectively.
5. The optimum value of P/e for discrete arc and V-shape ribs is 10 and 8, respectively, for low Reynolds number range, and for the higher range, the trend reverses. Similarly, the optimum values of W/w at a low Reynolds number range are observed at 5 and 6 for multiple arc and V-shape ribs, respectively. The same trends have been observed for discrete multiple arc and V-shape ribs. The difference in these optimum values is fairly significant, which necessitates in-depth investigations.
6. For discrete multiple rib configurations, the effective efficiency is higher for arc ribs as compared to V-shape ribs in the medium Reynolds number range. The reason for this trend is unidentified and requires in-depth fluid dynamics analyses.

This comparative investigation of different configurations of arc and V-shape ribs in solar air channel suggest that despite the similar design of arc and V-shape ribs, their thermo-hydraulic performance is significantly different. Apart from this, some deviations in optimum geometrical parameter values have also been identified. This study triggers the research toward finding the cause for such deviations in the performance of arc and V-shape rib configurations.

NOMENCLATURE

$h_{tc\leftrightarrow A}$	Convection heat transfer from transparent cover to ambience
$h_{tc\to sf}$	Convection heat transfer from transparent cover to stagnant fluid

$h_{R,\ tc \leftrightarrow ap}$	Radiative heat transfer from transparent cover to absorber plate
$h_{R,\ tc \leftrightarrow A}$	Radiative heat transfer from transparent cover to ambience
T_{tc}	The temperature of the transparent cover
T_A	Ambient air temperature
T_{sf}	Stagnant fluid temperature
T_{ap}	Absorber plate temperature
I_t	Total solar insolation
α_{tc}	Absorptivity of the transparent cover
σ	Stefan-Boltzmann constant
ε_{tc}	The emissivity of the transparent cover
E_{ap}	The emissivity of the absorber plate
$h_{ap \rightarrow sf}$	Convective heat transfer from absorber plate to stagnant fluid
$h_{ap \rightarrow f}$	Convective heat transfer from absorber plate to fluid in the channel
$h_{R,\ ap \leftrightarrow B}$	Radiative heat transfer from absorber plate to the bottom wall
T_f	Fluid temperature
T_B	Bottom wall temperature
α_{ap}	Absorptivity of the absorber plate
τ_{tc}	Transitivity of the transparent cover
Q_u	The useful heat gain by fluid
m_a	The mass flow rate of air
$c_{p,\ a}$	Specific heat of the air
T_{ai}	The temperature of the air at the inlet
T_{ao}	The temperature of the air at the outlet
$A_{f,\ ap}$	The frontal area of the absorber plate for solar radiation
$E_{n,\ o}$	The net energy output of the solar collector
P_{mech}	Mechanical power
CF	Conversion factor
ρ_a	The mass density of air
$(\Delta p)_c$	Pressure drop in channel
η_{th}	The thermal efficiency of solar collector
Lc	Length of channel
$\eta_{\text{effective}}$	Effective efficiency of solar collector
W_c	Width of channel
H_c	Height of channel
N_{tc}	Number of the transparent covers
g/e	Relative gap width
P/e	Relative roughness pitch
W/w	Relative roughness width
A	Angle of attack
d/w	Relative gap position
k_a	Thermal conductivity of air
e/D_h	Relative roughness height
μ_a	Dynamic viscosity of air
T_{am}	Mean air temperature
T_{apm}	Absorber plate mean temperature

$Q_{u,\,eb}$	Useful heat gain evaluated by energy balance
$Q_{u,\,hr}$	Useful heat gain evaluated by heat removal factor
U_{Total}	Total thermal loss
U_T	Thermal loss from the top surface
U_B	Thermal loss from the bottom wall
U_E	Thermal loss from edges
D_h	Hydraulic diameter
N_u	Nusselt number
F	Friction factor

SUBSCRIPTS

d_a	Discrete arc rib
d_V	Discrete V-shape rib
m_a	Multiple arc rib
m_V	Multiple V-shape rib
d_{ma}	Discrete multiple arc rib
d_{mV}	Discrete multiple V-shape rib

REFERENCES

[1] A. Awasthi, A. Mohan and A. Sharma, Tilt angle optimization and formulation of annual adjustment models for a solar power plant setup-able site at Himachal Pradesh, India, *IJITEE* 9 (2020), 1181–1185.

[2] A. Sharma, R. Chauhan, T. Singh, A. Kumar, R. Kumar, A. Kumar et al., Optimizing discrete V obstacle parameters using a novel Entropy-VIKOR approach in a solar air flow channel, *Renewable Energy* 106 (2017), 310–320.

[3] R. Chauhan and N.S. Thakur, Heat transfer and friction characteristics of impinging jet solar air heater, *Journal of Renewable and Sustainable Energy* 4 (2012), 043121.

[4] T. Josyula, S. Singh and P. Dhiman, Numerical investigation of a solar air heater comprising longitudinally finned absorber plate and thermal energy storage system, *Journal of Renewable and Sustainable Energy* 10 (2018), 055901.

[5] S. Nain, V. Ahlawat, S. Kajal, P. Anuradha, A. Sharma and T. Singh, Performance analysis of different U-shaped heat exchangers in parabolic trough solar collector for air heating applications, *Case Studies in Thermal Engineering* 25 (2021), 100949.

[6] A. Sharma, M.A. Kallioğlu, A. Awasthi, R. Chauhan, G. Fekete and T. Singh, Correlation formulation for optimum tilt angle for maximizing the solar radiation on solar collector in the Western Himalayan region, *Case Studies in Thermal Engineering* 26 (2021), 101185.

[7] A. Sharma, R. Chauhan, M. Ali Kallioğlu, V. Chinnasamy and T. Singh, A review of phase change materials (PCMs) for thermal storage in solar air heating systems, *Materials Today: Proceedings* 44 (2021), 4357–4363.

[8] S.K. Saini and R.P. Saini, Development of correlations for Nusselt number and friction factor for solar air heater with roughened duct having arc-shaped wire as artificial roughness, *Solar Energy* 82 (2008), 1118–1130.

[9] V.S. Hans, R.S. Gill and S. Singh, Heat transfer and friction factor correlations for a solar air heater duct roughened artificially with broken arc ribs, *Experimental Thermal and Fluid Science* 80 (2017), 77–89.

[10] A.P. Singh, Varun and Siddhartha, Heat transfer and friction factor correlations for multiple arc shape roughness elements on the absorber plate used in solar air heaters, *Experimental Thermal and Fluid Science* 54 (2014), 117–126.
[11] N.K. Pandey, V.K. Bajpai and Varun, Experimental investigation of heat transfer augmentation using multiple arcs with gap on absorber plate of solar air heater, *Solar Energy* 134 (2016), 314–326.
[12] A.M. Ebrahim Momin, J.S. Saini and S.C. Solanki, Heat transfer and friction in solar air heater duct with V-shaped rib roughness on absorber plate, *International Journal of Heat and Mass Transfer* 45 (2002), 3383–3396.
[13] S. Singh, S. Chander and J.S. Saini, Heat transfer and friction factor correlations of solar air heater ducts artificially roughened with discrete V-down ribs, *Energy* 36 (2011), 5053–5064.
[14] S. Singh, S. Chander and J.S. Saini, Heat transfer and friction factor of discrete V-down rib roughened solar air heater ducts, *Journal of Renewable and Sustainable Energy* 3 (2011), 013108.
[15] S. Singh, S. Chander and J.S. Saini, Thermal and effective efficiency based analysis of discrete V-down rib-roughened solar air heaters, *Journal of Renewable and Sustainable Energy* 3 (2011), 023107.
[16] V.S. Hans, R.P. Saini and J.S. Saini, Heat transfer and friction factor correlations for a solar air heater duct roughened artificially with multiple v-ribs, *Solar Energy* 84 (2010), 898–911.
[17] A. Kumar, R.P. Saini and J.S. Saini, Development of correlations for Nusselt number and friction factor for solar air heater with roughened duct having multi v-shaped with gap rib as artificial roughness, *Renewable Energy* 58 (2013), 151–163.
[18] D. Wang, J. Liu, Y. Liu, Y. Wang, B. Li and J. Liu, Evaluation of the performance of an improved solar air heater with "S" shaped ribs with gap, *Solar Energy* 195 (2020), 89–101.
[19] P.K. Jain and A. Lanjewar, Overview of V-RIB geometries in solar air heater and performance evaluation of a new V-RIB geometry, *Renewable Energy* 133 (2019), 77–90.
[20] A. Kumar and A. Layek, Nusselt number and friction factor correlation of solar air heater having twisted-rib roughness on absorber plate, *Renewable Energy* 130 (2019), 687–699.
[21] T.T. Ngo and N.M. Phu, Computational fluid dynamics analysis of the heat transfer and pressure drop of solar air heater with conic-curve profile ribs, *Journal of Thermal Analysis and Calorimetry* 139 (2020), 3235–3246.
[22] Y.M. Patel, S.V. Jain and V.J. Lakhera, Thermo-hydraulic performance analysis of a solar air heater roughened with reverse NACA profile ribs, *Applied Thermal Engineering* 170 (2020), 114940.
[23] S. Thakur and N.S. Thakur, Impact of multi-staggered rib parameters of the 'W' shaped roughness on the performance of a solar air heater channel, *Energy Sources, Part A: Recovery, Utilization, and Environmental Effects* (2020), 1–20. doi: 10.1080/15567036.2020.1764672.
[24] S. Thakur and N.S. Thakur, Investigational analysis of roughened solar air heater channel having W-shaped ribs with symmetrical gaps along with staggered ribs, *Energy Sources, Part A: Recovery, Utilization, and Environmental Effects* (2019), 1–16. doi: 10.1080/15567036.2019.1675815.
[25] S.K. Jain, G.D. Agrawal and R. Misra, Heat transfer augmentation using multiple gaps in arc-shaped ribs roughened solar air heater: An experimental study, *Energy Sources, Part A: Recovery, Utilization, and Environmental Effects* (2019), 1–12. doi: 10.1080/15567036.2019.1607945.
[26] R. Chauhan and S.C. Kim, Effective efficiency distribution characteristics in protruded/dimpled-arc plate solar thermal collector, *Renewable Energy* 138 (2019), 955–963.

[27] A. Priyam and P. Chand, Thermal and thermohydraulic performance of wavy finned absorber solar air heater, *Solar Energy* 130 (2016), 250–259.

[28] T. Istanto, D. Danardono, I. Yaningsih and A.T. Wijayanta, Experimental study of heat transfer enhancement in solar air heater with different angle of attack of V-down continuous ribs, *AIP Conference Proceedings* 1737 (2016), 060002.

[29] A. Lanjewar, J.L. Bhagoria and R.M. Sarviya, Experimental study of augmented heat transfer and friction in solar air heater with different orientations of W-Rib roughness, *Experimental Thermal and Fluid Science* 35 (2011), 986–995.

[30] I. Singh and S. Singh, CFD analysis of solar air heater duct having square wave profiled transverse ribs as roughness elements, *Solar Energy* 162 (2018), 442–453.

5 Lithium-Based Batteries Charged by Regenerative Braking Using Second Quadrant Chopper

Mohit Garg, Naresh Kumar, and Anil Kumar
DCR University of Science & Technology

CONTENTS

5.1 INTRODUCTION

In the coming future, the scope of emerging technologies will be quite large. As there is a lot of development going on nowadays, some significant areas are lagging behind. Conventional vehicles, which are based on IC engines, are one of the major areas [1]. These vehicles are even improving in terms of producing pollution, but the pollution produced even now produces much toxic gases such as SO_2 [2]. Government institutions and various national and international institutions are showing their efforts in this respect [3]. So, in order to resolve these issues, the research in the area of EVs is increasing at a high rate. So, there are not sufficient, but some EVs, particularly e-rickshaws, available [4]. There are various reasons for their less development, such as charging stations, structure for lightweight electric vehicles, charging and discharging process, and the efficiency of the battery [5]. But as we can see most of these factors are related to battery only. So, in this chapter, the research is based on lithium batteries using regenerative braking, one of the techniques of battery developments [6].

In this era of technology, where there is progress in every field, the battery development field is even lagging behind now [7]. The area of research is also too vast

DOI: 10.1201/9781003272717-5

in this field, but due to slow progress, we are facing problems such as high cost. The battery in the EVs takes almost 45%–50% cost of the EVs [8]. So, today, there is large research going on battery optimization techniques. Today, the focus is on lithium-ion-based batteries with high energy and power density, low self-discharge, and high efficiency [9]. But besides of all the advantages, these batteries also have major disadvantages such as high cost, lithium plating at low temperature, sensitivity to temperature, and explosion due to overcharge [10]. So, maintaining all of these factors for lithium-ion-based batteries requires implementing battery optimization techniques [11]. One of the battery optimization techniques is regenerative braking. Regenerative braking achieves an increase in the state of charge (SoC) of the battery by recovering the vehicle's kinetic energy during the vehicle braking. In city areas, where there is a problem of traffic always, most of the battery's energy gets consumed in braking [12]. Regenerative braking can be proved to be useful in these situations to charge the battery using this kinetic energy of the vehicle, and in this way, it can increase the range of the vehicle and can also increase the life of the battery [13]. The flywheel can also be implemented in this system to increase the effectiveness, but it suffers from various limitations of space, mechanical strength, etc. [14].

Some papers are there on the regenerative braking control system. The predictive controller can achieve the recovery of energy with a non-linear model [15]. Using fuzzy control, the displacement of the brake pedal, variation rate of the brake pedal, vehicle speed, and wheel slip rate error are used as inputs [16]. Criteria in paper [17] took vehicle speed and the desired force of motor as the input. However, the above research did not consider the SOC of energy storage devices, which is a significant point that affects energy recovery efficiency, and it may cause damage to the battery [18]. Most of the papers use an inverter as a controller to achieve the strategy of recovering the energy.

This study uses the chopper as a controller to recover the energy while braking. It uses the second quadrant operation of the chopper to recover the energy; i.e., the second quadrant chopper is a regenerating mode of the quadrant where the forward braking takes place [19]. It also uses a closed-loop control system, where the reference speed, the actual speed of the dc motor, and the armature current of the dc motor are taken as inputs to the subsystem of the closed-loop system.

5.2 DESIGNING AND WORKING OF THE CONVERTER

The converter used here is the second quadrant chopper. This quadrant shows the operation of power transfer from the output side to the input side; i.e., it shows the mode of regenerative braking in the case of motors. At the secondary end, i.e., load side, the DC motor is connected, which shows that the motor which we are trying to use in EV is a DC motor [20]. Various DC motors can be used, such as brushed DC motor, brushless DC motor, and synchronous motor [21–23]. But here, only the simple DC motor is used to show the effect of both the control system and to compare them easily. The design is done so that the Li-ion battery is connected at the input side, while at the output side, the DC motor is connected. The motor used is actually separately excited [24]. So, for the field supply, a

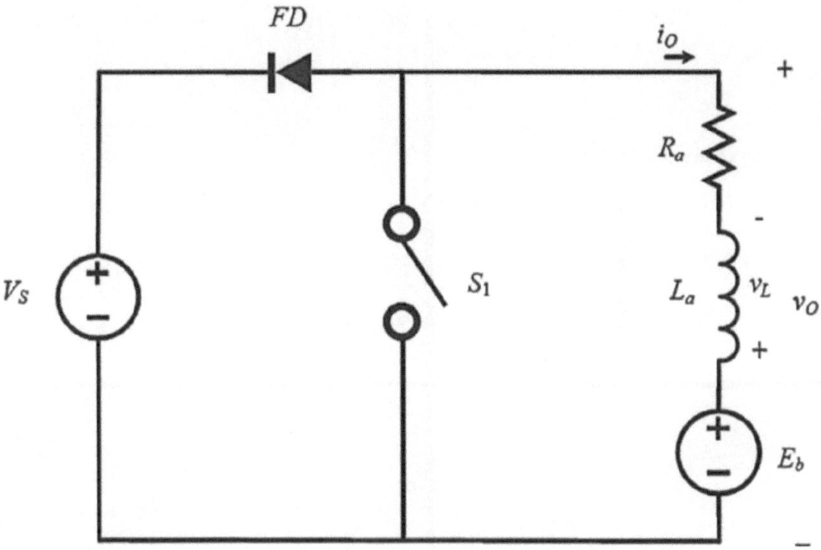

FIGURE 5.1 Schematic view of second quadrant chopper.

separate dc source is used. Figure 5.1 shows the schematic of the second quadrant chopper.

5.2.1 V_o–I_o Plane and Operation of Model

It works in the second quadrant, which is also shown in Figure 5.2, so, in this quadrant, it works as a generator [25]. In the first quadrant, the motor takes power from the source (battery used here), while in the second quadrant, the same motor will start releasing its kinetic energy, which gets stored in the inductor first and then will be supplied to the battery. In forward braking, the torque becomes negative, while the speed remains positive. So, the power becomes negative using the product of torque and speed involved [26,27].

Below, the two modes of operation are explained in brief:

Mode 1: In Mode 1, Switch S1 gets closed or we can say that it becomes ON. After closing the switch, the circuit becomes something similar to that shown below in Figure 5.3. The closed-loop shown in Figure 5.3 becomes the only conducting path.

The current starts flowing in a negative direction compared to the forward motoring mode of operation in the first quadrant. Now, the negative current makes the torque also negative.

$$\text{The energy released by motor} = 0.5\,\text{Jw}^2$$

$$\text{which gets stored in inductor as} = 0.5\,\text{Li}^2$$

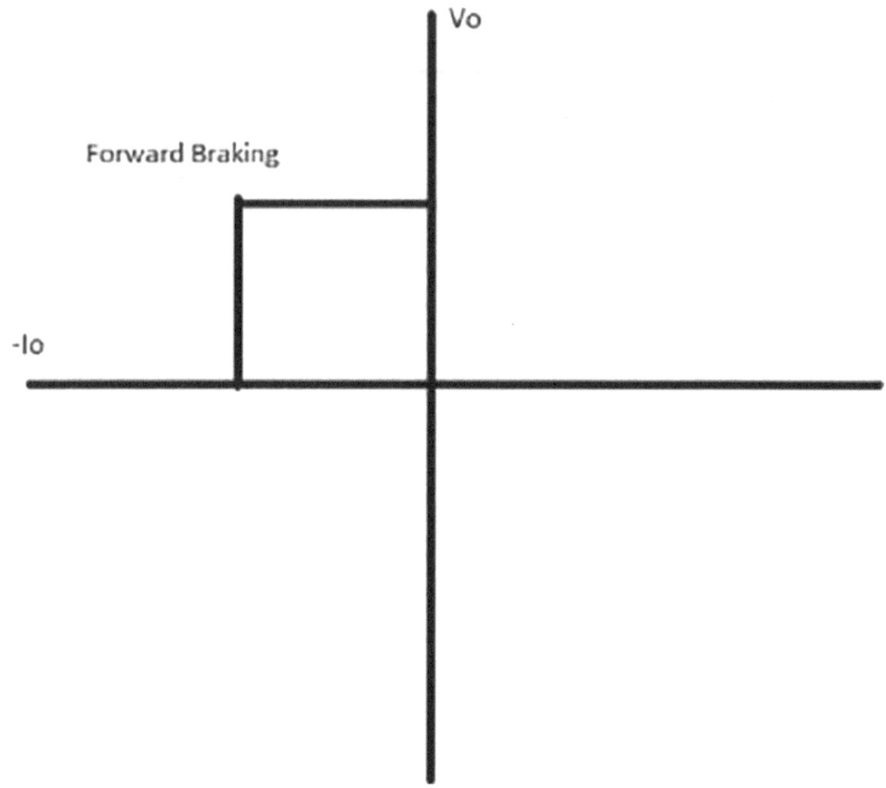

FIGURE 5.2 V_o–I_o plane of chopper.

In this way, the motor releases its energy and the inductor stores it in the form of magnetic energy.

Mode 2: In Mode 2, Switch S1 becomes OFF. After turning OFF the switch, the circuit becomes something similar to that shown below in Figure 5.4. The closed-loop shown in Figure 5.4 becomes the only conducting path. In this way, the energy stored in inductor in Mode 1, i.e.,

$$\alpha = 0.5\mathrm{Li}^2$$

will now be stored in the battery.

5.2.1.1 Mathematical Analysis

The output voltage of the circuit is given by:

$$V_o = V_s\left(\frac{T_{off}}{T}\right) \tag{5.1}$$

where

$$\frac{T_{off}}{T} = 1 - \alpha \tag{5.2}$$

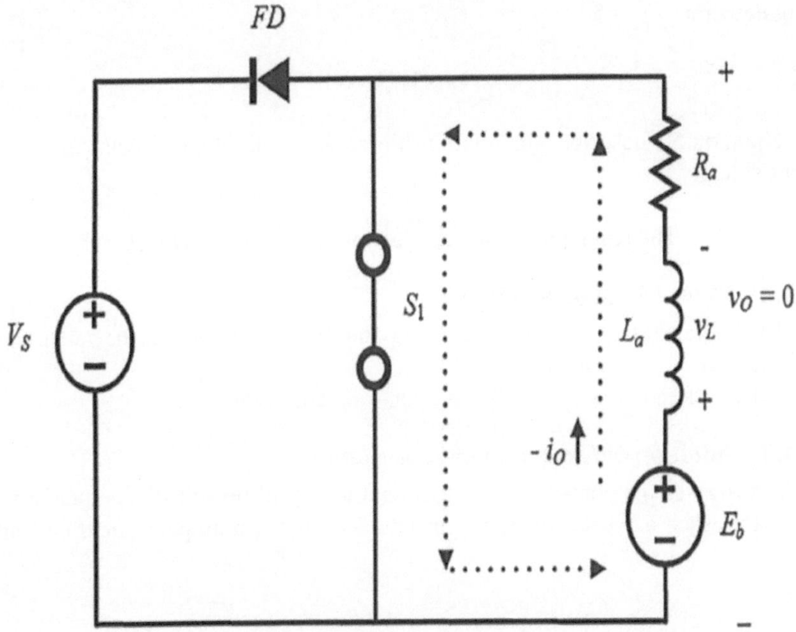

FIGURE 5.3 Mode 1 of second quadrant chopper.

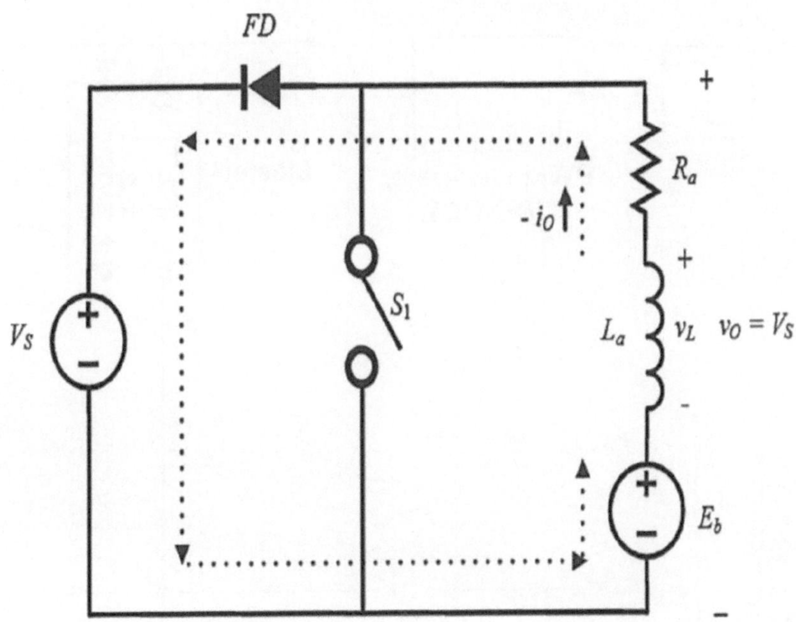

FIGURE 5.4 Mode 2 of second quadrant chopper.

Using equation 5.2 in 5.1,

$$V_o = V_s(1-\alpha) \tag{5.3}$$

This equation 5.3 describes the relationship between the supply voltage and average output voltage.

The regenerative power is given by: $V_o I_o = (1-\alpha)V_s I_o$ (5.4)

5.2.1.2 PWM Signal to MOSFET

The PWM generator is used here using the step function for the switching of MOSFET, as shown in Figure 5.5.

So, the system made is actually an open-loop controller.

5.2.1.3 Addition of Closed-Loop Subsystem

In this design of the controller, an additional loop will be added. The performance parameters such as speed (in rpm), armature current (in ampere), field current (in

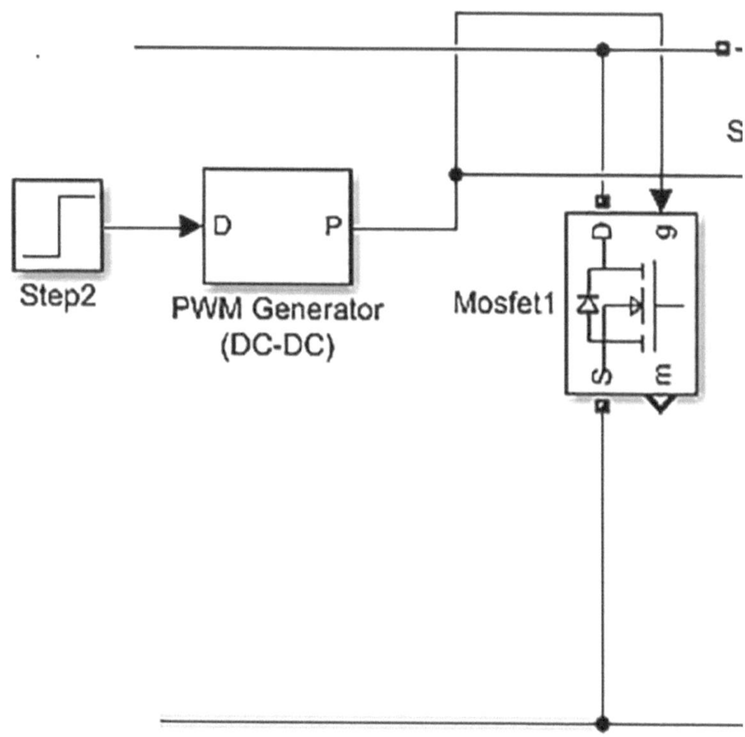

FIGURE 5.5 PWM signal to MOSFET.

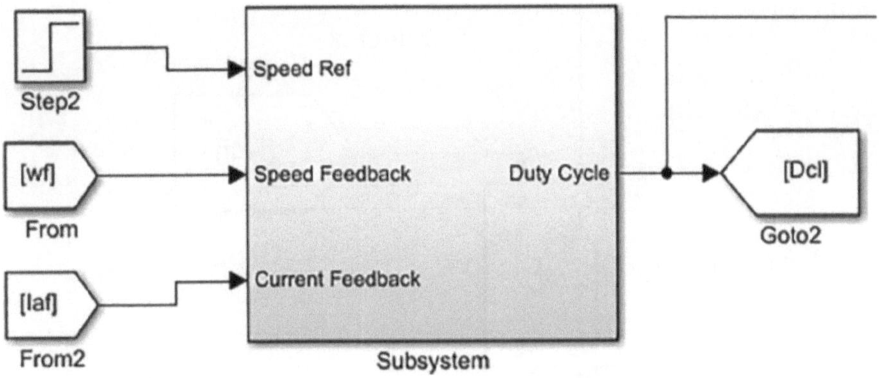

FIGURE 5.6 Subsystem of closed-loop control system for regenerative braking using second quadrant chopper as converter.

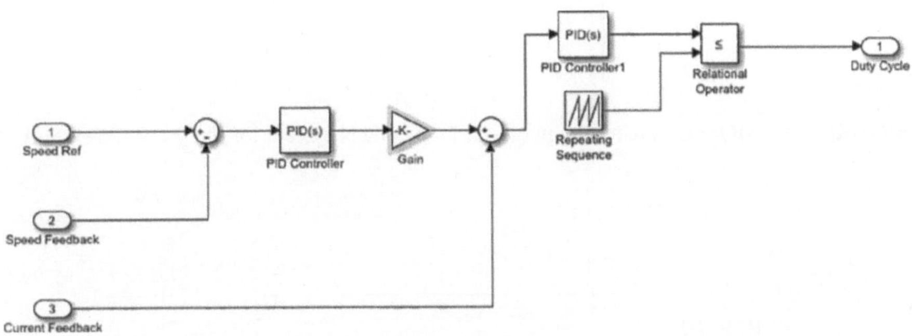

FIGURE 5.7 Explored components of the subsystem used.

ampere), torque (in N-m) are fed back. These parameters are fed back to a subsystem, as shown in Figure 5.6 [28].

The explored subsystem is shown in Figure 5.7.

The value of gain is adjusted at 0.76277. The output of subsystem, which is from Goto block Goto 2, is sent to the MOSFET as shown in Figure 5.8.

So, in this way, the control system is designed for the closed-loop control system (Table 5.1).

5.3 SIMULATION RESULTS

The simulation for the proposed system is performed in MATLAB Simulink. The components for the simulation of the proposed system are taken from the Simulink library browser [29]. A DC motor of 5 HP is taken. The DC motor selected is of separately excited type. That's why, for its performance, a separate source is provided. The Li-ion battery is taken here with 50% SoC. The switching in the circuit is provided through the MOSFET [30] switch, which is controlled using the output of feedback control transfer function by the subsystem used, as shown in Figure 5.6.

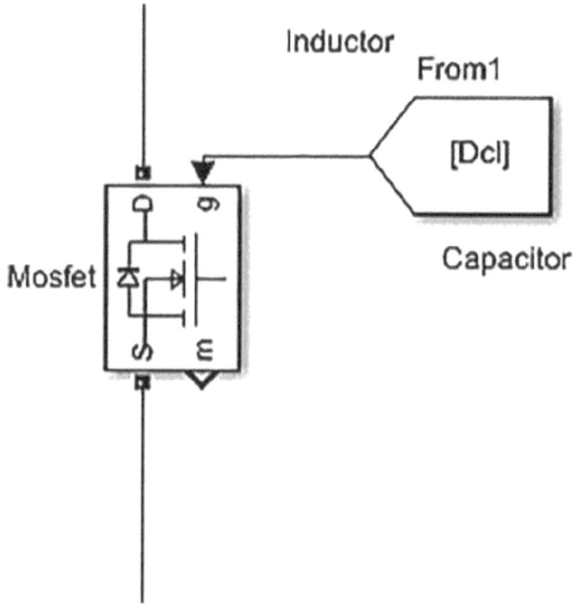

FIGURE 5.8 MOSFET switching in closed-loop control system for the proposed system.

TABLE 5.1
Parameters and Their Values of Components Used in Regenerative Braking

Parameters	Values
Battery initial SoC	50%
Battery nominal voltage	240 V
Battery rated capacity	20 Ah
DC motor power	5 HP, 1750 rpm
DC motor RPM	1750 rpm
DC motor voltage rating	240 V
Initial value of negative torque	−20 N-m
Final value of negative torque	−15 N-m
Gain value (K)	0.76277
Diode resistance	0.001 ohm
Diode forward voltage	0.8 V
Diode snubber resistance	500 ohm
Diode snubber capacitance	250×10^{-9} F
Filter inductance	5 mH
Filter capacitance	150×10^{-6} F

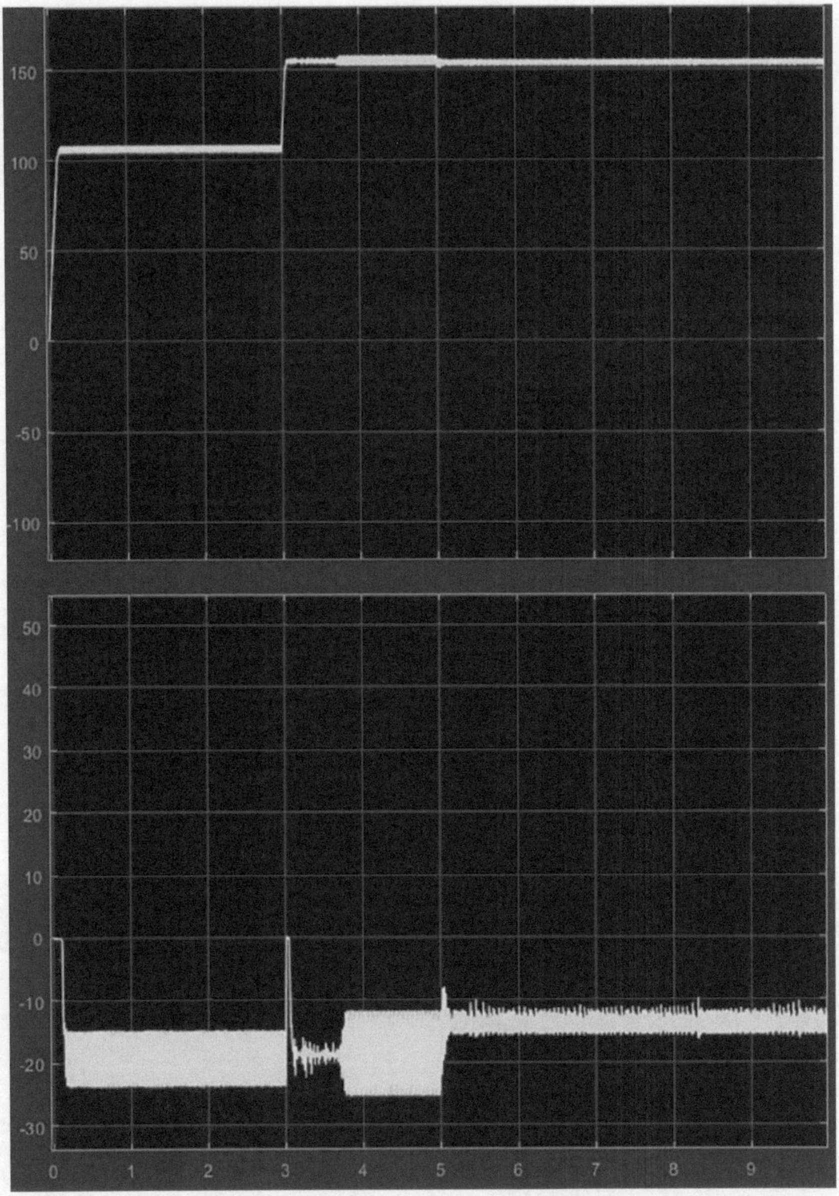

FIGURE 5.9 Graphs of speed and torque.

Figure 5.9 shows the result in the form of a graph describing various types of current and voltage in the circuit. They describe the power and SoC of the battery also (Figure 5.10). The explanation of the result in this case of regenerative braking using the second quadrant like this is that at the time of 3 seconds when the change in current is from 100 to 150 rpm as shown in Figure 5.9, then there is an increase in the power flow in the circuit as shown in Figure 5.11.

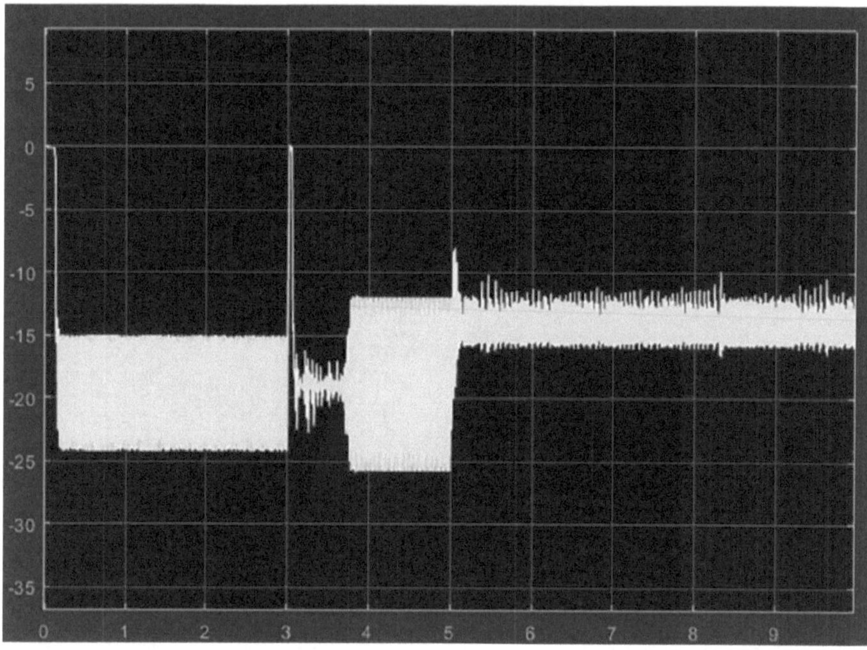

FIGURE 5.10 Graph of armature current.

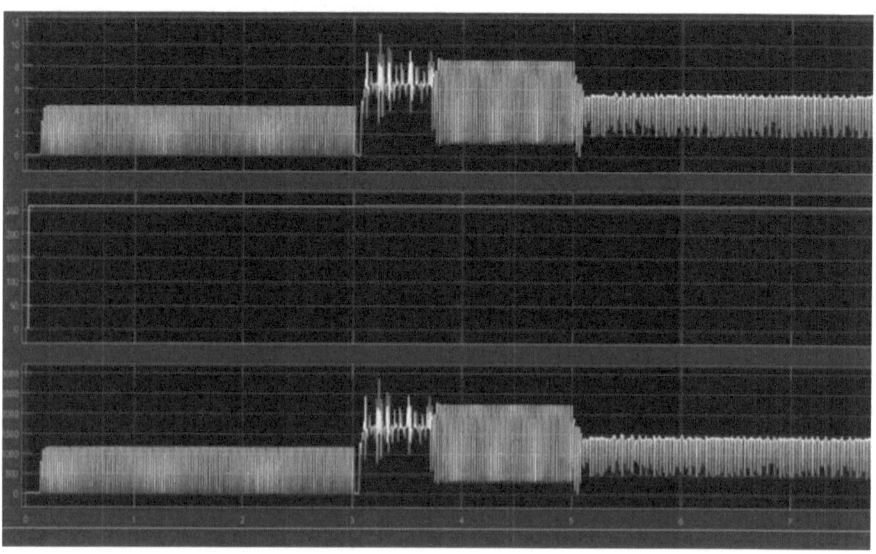

FIGURE 5.11 Graphs of current, voltage, and power at the battery terminals of the model.

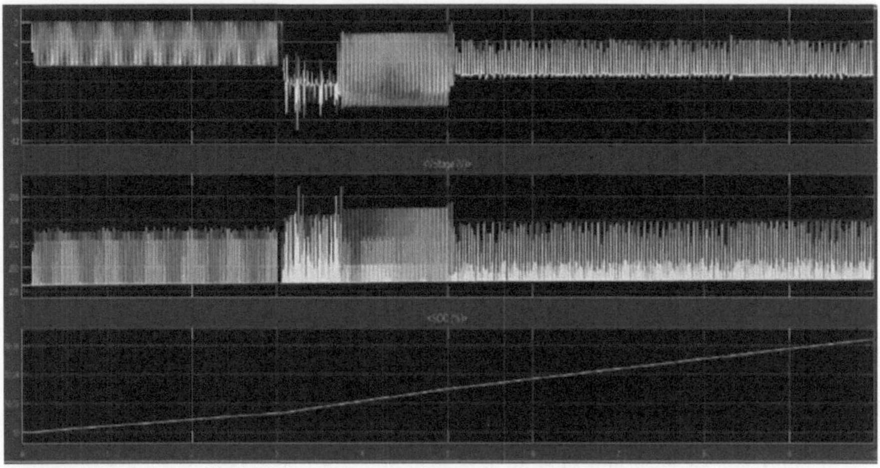

FIGURE 5.12 Graphs of current, voltage, and SoC in the battery.

But here, the voltage across the input terminals of the circuitry, i.e., at battery terminals, remains the same while there is an increase in the current flowing at the battery terminals side to show the increase in power with an increase in speed. The increase in speed can indicate the factor on slope-type land.

Now, as there is an increase in power in circuitry, so there will be an increase in the charging of the battery to simulate that as there is an increase in speed while on sloppy areas or due to some factors may be, then there would be an increase in the SoC of the battery. And that's why there is an increase in the slope of the SoC of the battery as shown in Figure 5.12.

5.4 CONCLUSIONS

Regenerative braking using a second quadrant chopper has been proposed in this chapter. In this chapter, the battery is charged with the kinetic energy produced during the braking of the electric vehicle. The energy that was previously lost due to the friction between wheel and land is now used to charge the battery. It necessitates only one switch. The circuit proposed is modeled in MATLAB using a 5 HP, 240 V, 1750 rpm, separately excited dc motor and a 20 Ah lithium-ion battery. The proposed technique is fast in operation and also quite economical and straightforward. It is easy to implement. It helps in improving the efficiency, lessening charging necessities, and increasing the vehicle range.

REFERENCES

[1] S. Schiffer, A. Kain, P. Wilde, M. Helbing and B. Baeker, *2017 Twelfth International Conference on Ecological Vehicles and Renewable Energies (EVER) – Improving Longitudinal Dynamics of Conventional Vehicles in Comparison to Electrified Vehicles to Meet Customer Behavior*, 2017.

[2] T. Peshin, I. M. L. Azevedo and S. Sengupta, Life-cycle greenhouse gas emissions of alternative and conventional fuel vehicles in India, *2020 IEEE Vehicle Power and Propulsion Conference (VPPC)*, 2020.
[3] G. Jungmeier, J. B. Dunn, A. Elgowainy, E. D. Ozdemir, S. Ehrenberger, H. J. Althaus and R. Widmer, *2013 World Electric Vehicle Symposium and Exhibition (EVS27) – Key Issues in Life Cycle Assessment of Electric Vehicles' Findings in the International Energy Agency (IEA) on Hybrid and Electric Vehicles (HEV)*, 2013.
[4] P. Nambisan, S. Bansal and M. Khanra, *2020 IEEE International Conference on Power Electronics, Smart Grid and Renewable Energy (PESGRE2020) – Economic Performance of Solar Assisted Battery and Supercapacitor based E-Rickshaw*, 2020.
[5] Y. Cao et al., An optimized EV charging model considering TOU price and SoC curve, *IEEE Transactions on Smart Grid*, 2012, 3, 1, pp. 388–393.
[6] T. T. Duong, D. V. Do, T. T. Nguyen, *2018 4th International Conference on Green Technology and Sustainable Development (GTSD) – Research on Braking Force Distribution in Regenerative Braking System Apply to Conventional Vehicle*, 2018.
[7] W. Sutopo, R. Ardiansyah, Yuniaristanto, M. Nizam, *2013 Joint International Conference on Rural Information & Communication Technology and Electric-Vehicle Technology (rICT & ICeV-T) – An Application of Parametric Cost Estimation to Predict Cost of Electric Vehicle Prototype*, 2013.
[8] P. Prevedouros, L. Mitropoulos, *2018 21st International Conference on Intelligent Transportation Systems (ITSC) – Impact of Battery Performance on Total Cost of Ownership for Electric Drive Vehicle*, 2018.
[9] Samaras and K. Meisterling, Life cycle assessment of greenhouse gas emissions from plug-in hybrid vehicles: Implications for policy, *Environmental Science & Technology*, 2008, 42, pp. 3170–3176.
[10] M. Yi et al., Detection of lithium plating based on the distribution of relaxation times, *2021 IEEE 4th International Electrical and Energy Conference (CIEEC)*, 2021, pp. 1–5,
[11] D. Karunathilake, M. Vilathgamuwa, T. W. Yateendra Mishra, Farrell2, San Shing Choi, Capacity loss reduction using smart-battery management system for li-ion battery energy storage systems. *2020 IEEE 29th International Symposium on Industrial Electronics (ISIE)*, 2020.
[12] Y. Gao, L. Chen and M. Ehsani, Investigation of the effectiveness of regenerative braking for EV and HEV, *SAE International SP1466*. 1999-01-2910, 1999.
[13] H. Liu, X. He, L. Chu and J. Tian, Study on control strategy of regenerative braking for electric bus based on braking comfort, *Proceedings of 2011 International Conference on Electronic & Mechanical Engineering and Information Technology*, 2011.
[14] J. Dixon, Energy storage for electric vehicles. *Industrial Technology (ICIT) 2010 IEEE International Conference*, 2010, pp. 20–26.
[15] X. Huang and J. Wang, Model predictive regenerative braking control for light weight electric vehicles with in-wheel motors, *Journal of Automobile Engineering*, 2012, 226, 9, pp. 1220–1232.
[16] L. Chu, J. Yin, L. Yao and W. Wang, Study on the braking force allocation dynamic control strategy based on the fuzzy control logic, *Industrial Engineering and Engineering Management (IE&EM), 2011 IEEE 18Th International Conference on*, 2011, 1, pp. 635–636.
[17] Z. Zhang, W. Li, G. Xu and L. Zheng, Regenerative braking for electric vehicle based on fuzzy logic control strategy, *2010 2nd International Conference on Mechanical and Electronics Engineering*, 2010, 1, pp. 319–320.
[18] C. Li, C. He, Y. Yuan and J. Zhang, Braking evaluation of integrated electronic hydraulic brake system equipped in electric vehicle, *2019 IEEE 3rd Information Technology, Networking, Electronic and Automation Control Conference (ITNEC)*, 2019, pp. 2361–2365.

[19] T. Umihara, A. Tamura, T. Ishibashi and A. Kawamura, Proposal of soft SOC balancing method to two battery HEECS chopper used for EV power train, *IECON 2018–44th Annual Conference of the IEEE Industrial Electronics Society*, 2018, pp. 2128–2132.
[20] A. Prince, P. K. Abraham and B. Aryanandiny, Design and implementation of ultra capacitor based regenerative braking system for a DC motor, *2018 International Conference on Emerging Trends and Innovations in Engineering and Technological Research (ICETIETR)*, 2018, pp. 1–4.
[21] R. G. Chougale and C. R. Lakade, Regenerative braking system of electric vehicle driven by brushless DC motor using fuzzy logic, *2017 IEEE International Conference on Power, Control, Signals and Instrumentation Engineering (ICPCSI)*, 2017, pp. 2167–2171.
[22] S. M. B. Billah and K. K. Islam, Regenerative braking characteristics of PMDC motor by applying different armature voltage, *2016 2nd International Conference on Electrical, Computer & Telecommunication Engineering (ICECTE)*, 2016, pp. 1–4.
[23] O. Sinchuk and I. Kozakevich, Research of regenerative braking of traction permanent magnet synchronous motors, *2017 International Conference on Modern Electrical and Energy Systems (MEES)*, 2017, pp. 92–95.
[24] A. T. Waghe and S. K. Patil, A bidirectional DC-DC converter fed separately excited DC motor electric vehicle application, *2020 International Conference on Emerging Trends in Information Technology and Engineering (ic-ETITE)*, 2020, pp. 1–5.
[25] K. Hirachi, M. Kurokawa and M. Nakaoka, Feasible compact UPS incorporating current-mode controlled two-quadrant chopper-fed battery link, *Proceedings of Second International Conference on Power Electronics and Drive Systems*, 1997, 1, pp. 418–424.
[26] Y. Saleem and T. Izhar, Control of torque in switched reluctance motor, *2008 Second International Conference on Electrical Engineering*, 2008, pp. 1–4
[27] R. L. Steigerwald, A two-quadrant transistor chopper for an electric vehicle drive, In *IEEE Transactions on Industry Applications*, 1980, IA-16, 4, pp. 535–541.
[28] G. Cohen, M. Sassonker and A. Kuperman, Closed-loop scalar speed control of induction motor drive with limited regenerative braking ability, *2021 IEEE 19th International Power Electronics and Motion Control Conference (PEMC)*, 2021, pp. 757–760
[29] N. Hinov, H. Kanchev and F. Krastev, A web-based MATLAB application for e-learning and simulation of electric vehicle performance, *2020 7th International Conference on Energy Efficiency and Agricultural Engineering (EE&AE)*, 2020, pp. 1–4
[30] T. Halder, Selection of switching skills of the power MOSFET in the static converter, *2017 14th IEEE India Council International Conference (INDICON)*, 2017, pp. 1–6.

6 Modeling and Simulation of SoC-Based BMS for Stand-Alone Solar PV-Fed DC Microgrids

Arambakam Sreeram, Yugal Kishor, C.H. Kamesh Rao, and R.N. Patel
National Institute of Technology

CONTENTS

DOI: 10.1201/9781003272717-6

6.1 DC MICROGRID ARCHITECTURE

6.1.1 DC Home

In India, the lack of electric power in rural areas and remote areas like tribal areas in forests of Nallamala and deserts of Rajasthan is mainly because of the geographical constraints and the economic problems [1–3].

A solution to the existing grid problems is employing DC microgrids, as explained with real-time analysis. The load demand (AC and DC) is met by a reliable and safe DC grid voltage profile with solar PV panel and ESS (batteries) as sources using appropriate converters, as shown in Figure 6.1. DC microgrids have several DC homes interconnected to enhance the microgrid's quality, consistency, and independent power [4].

A DC home consists of:

a. Rooftop PV Panel Installation: The sizing of the rooftop solar panel installation is done to meet the power demand. In the case of an interconnected DC microgrid system using DC home configurations, by clustering them together, the rooftop panel rating is chosen to be more than the individual demand such that the total generation and total load demand are met [5,6].
b. MPPT Charge Controller: To control the boost converter to reach the DC bus voltage level, a solar charging controller is used to implement the MPPT algorithm using the perturb and observe method. In the system used to analyze the proposed system, a 48 V DC bus voltage is selected, and a four-string PV array, each string having two panels, leads to a total of 72 V. A buck converter is needed to bring down the PV output to the bus voltage level.
c. DC Bus: Household appliances such as lights and fans and electronic devices such as televisions, computers, and mobile phones operate on DC supply; high power rating equipment such as induction stoves, mixers, washing machines, and refrigerators using DC supply is available in the market. Hence, the transmission from conventional AC distribution system to DC-based power distribution strategy is efficient and essential. The DC bus voltage is to be maintained at 48 V_{dc}.
d. BESS: A DC home configuration of PV-fed DC microgrids is a stand-alone application or works in islanded mode of operation. Since solar power can be harvested only in the daytime, a battery energy storage system is essential as the primary source for power backup. Batteries are the best way to store electrical energy.

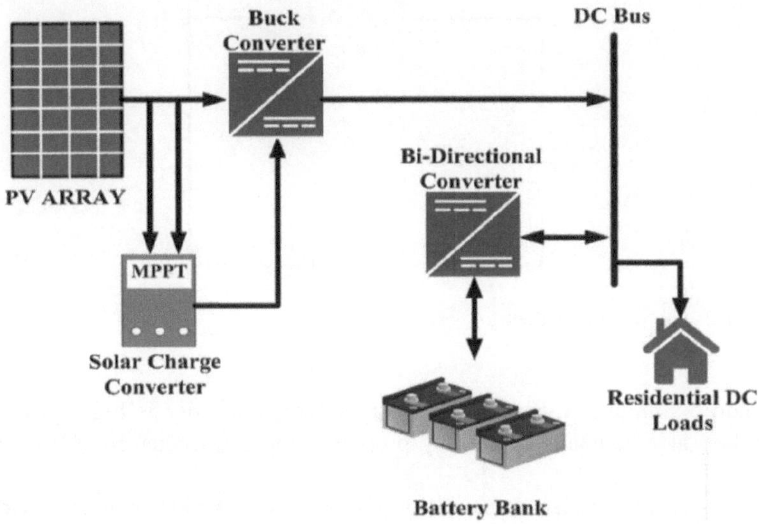

FIGURE 6.1 Power exchange diagram for an islanded DC microgrid.

TABLE 6.1
BESS Parameters

Battery Parameter	Value
Terminal voltage	24 V
State of charge (SoC)	10% < SoC < 98%
Battery capacity	50 Ah

e. BMS: The battery bank is to be maintained and used efficiently. A battery management system is the control algorithm designed for this sole purpose. The battery management comprises a charge and discharge algorithm and a battery scheduling algorithm to cut off a battery that's entering the deep discharge region.
f. DC Loads: All household electronic appliances work on DC power. Recent advancements in power electronics and choppers (DC/DC converters) appliances such as ACs, refrigerators, and washing machines are also popular.

6.1.2 Battery Energy Storage System (BESS)

A series-connected BESS is made of two batteries. For system design, a lead-acid battery is used. Battery parameters, i.e., state of charge (SoC) and terminal voltage of the battery, are shown in Table 6.1.

The overall battery bank terminal voltage will be 48 V (24 + 24), and the initial state of charge is 50% on average, whereas the battery bank capacity is 100 Ah (50 Ah each). The equivalent circuit battery model is popularly implemented

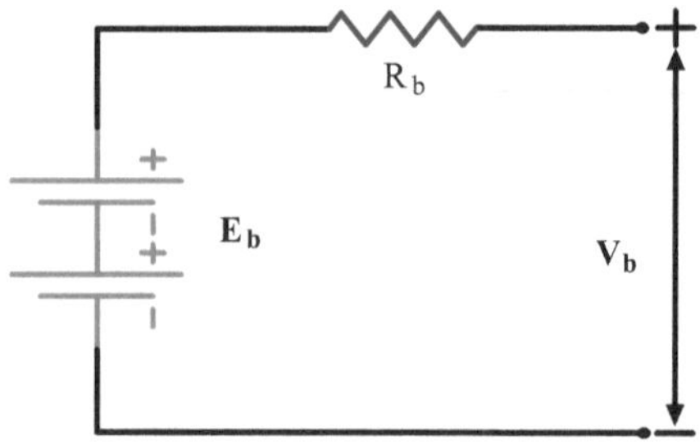

FIGURE 6.2 Battery equivalent.

for dynamic simulation. Since the battery's energy density is high, power can be supplied at a constant voltage by properly controlling charging/discharging cycles.

The low-cost, large-size accessibility of a plum-acid battery is more convenient for renewable systems. For dynamic model simulation, the generic lead-acid battery model is made of a dependent voltage source with series resistance. The equivalent battery circuit in Figure 6.2 has a constant current control system designed for power flow from and to battery bank. The control input E_b is associated with the current, the state of charge (SoC) of the battery, and the battery terminal voltage (V_b), as in [7].

$$V_{\text{bus}} = E_b - R_o \bullet i_b \tag{6.1}$$

For discharging:

$$V_b = E_b - P\frac{C}{C - i_b t} \bullet i_b t - P\frac{C}{C - i_b t} \bullet i_b^{\,*} + e^t \tag{6.2}$$

For charging:

$$V_b = E - P\frac{C}{C - i_b t} \bullet it - P\frac{C}{i_b t - 0.1.C} \bullet i_b^{\,*} + e^t \tag{6.3}$$

where i_b is the charging and discharging current (A), R_b is the individual battery internal resistance (Ω), E_o is the open-circuit (OC) voltage (V), P is the potential of polarization (V), and C is the battery charge capacity (Ah).

6.1.3 Addressing Economic Constraints of a DC Microgrid

Batteries are an essential part of a PV-fed DC microgrid. Li-ion and lead-acid batteries are mainly used for energy storage. These batteries need periodic maintenance, or in worst cases, a replacement will be done. This battery maintenance cost can be reduced by implementing an effective control algorithm to maximize battery lifetime [4,8].

The factors that distinguish batteries from typical electricity sources and underlie the modeling of the BESS are the following:

a. The capacity of the battery depends on the current discharge. The energy efficiency of a BESS from stored chemical energy to electricity is reduced at higher currents. The capacity of the battery is gradually different from the nominal value for higher load currents.
b. If a battery is idle, some of its usable capacity can be recovered. Due to electrochemical phenomena, time intervals when no current is drawn increase the power supply capacity and battery amount [9,10].

A battery operated in the deep discharge region results in high currents. This issue is not addressed in traditional DC homes, which led to increased maintenance costs and high installation costs, leading to unsatisfied economic operation [10]. For a multi-battery system implementing the second property, an energy management scheme is developed. In this proposed scheme, the load is shared among the batteries in the battery bank, and hence, those idle batteries can recover during the idle period.

6.1.4 Analyzing the Optimization Problem in the Battery Storage System

The issue that has been addressed is deciding which cell (battery pack) should supply the load demand in the installed multi-battery system. The overall lifetime of the system reaches the ideal case (monolithic battery). The current load profile and BESS configuration are factors that significantly impact the efficiency of implemented BMS strategy [7,11]. To address this problem, we follow one of the three sequential discharging algorithms, dynamic switching, where each cell is discharged concerning its capacity by cutting off those cells whose further utilization will lead to deep discharge operation from the battery bank. In the below section, the dynamic switching algorithm is discussed in detail. A bidirectional DC/AC converter proposed in [12] is used if the DC home (DC microgrid) is provided with AC grid connectivity.

6.2 BATTERY MANAGEMENT SYSTEM (BMS)

Need for BMS:

The last 4%–10% battery region is regarded as the deep discharge region, as shown in Figure 6.3. In the deep discharge region, as the charge decreases, the voltage offered by the battery at its terminals will also decline drastically.

In cases like this, BESS supplies more current to meet the load power demands to compensate for the reduced voltage to maintain power. These high currents will be dangerous enough to cause the insulation to break down the wiring and damage the connected loads [13].

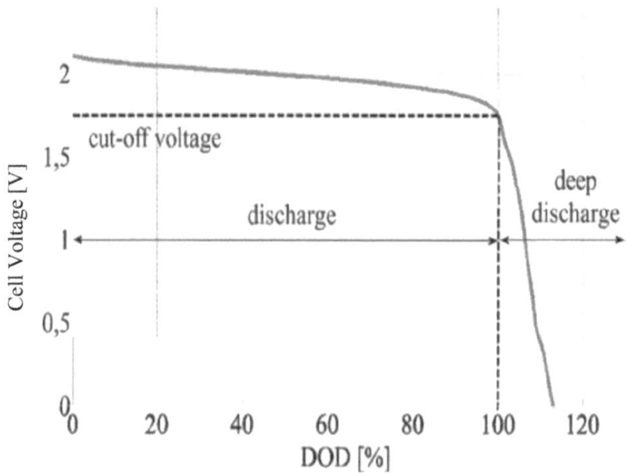

FIGURE 6.3 Discharge characteristics of lead-acid battery.

Once a cell of BESS enters deep discharge, its state of health (SoH) declines. Continuous operation of a battery in the deep discharge region can lead to battery breakdown. The BMS of a battery is an intelligent control algorithm implemented for optimal utilization and safe operation of the battery such that its lifetime is enhanced [14].

In the designed model of DC home, the proposed BMS comprises two control strategies, namely:

a. Battery charge/discharge control scheme.
b. Battery scheduling.

6.2.1 BESS Charge/Discharge Control Scheme

The implementation of switched charging/unloading enhances the planned battery time. SoC is seen as the factor determining the direction of power flow between the DC bus and the battery bank in the control strategy proposed. Every time the PV is in the battery bank, the SoC level is always charged up to 90%. Once it reaches the set maximum SoC level, the battery ON/OFF switch will be opened, and the battery bank is isolated from the system [15,16].

Similarly, when PV is OFF, the loads connected to the DC bus are supplied by the battery bank; i.e., the battery will be discharging. The minimum SoC level to which the battery bank can discharge is considered at 20%. Once it reaches the set minimum level, the battery bank will discharge through the emergency line until SoC reaches 5%, whereas BESS will be cut off from the bus. Thus, the BESS charge and discharge control strategy is developed, and the implementation is shown in Figure 6.4. In the simulations of DC home configuration, the same control strategy is implemented. SoC limits of 10% and 98%, respectively, are set at minimum and maximum. Figure 6.5 shows the execution of the algorithm above [17].

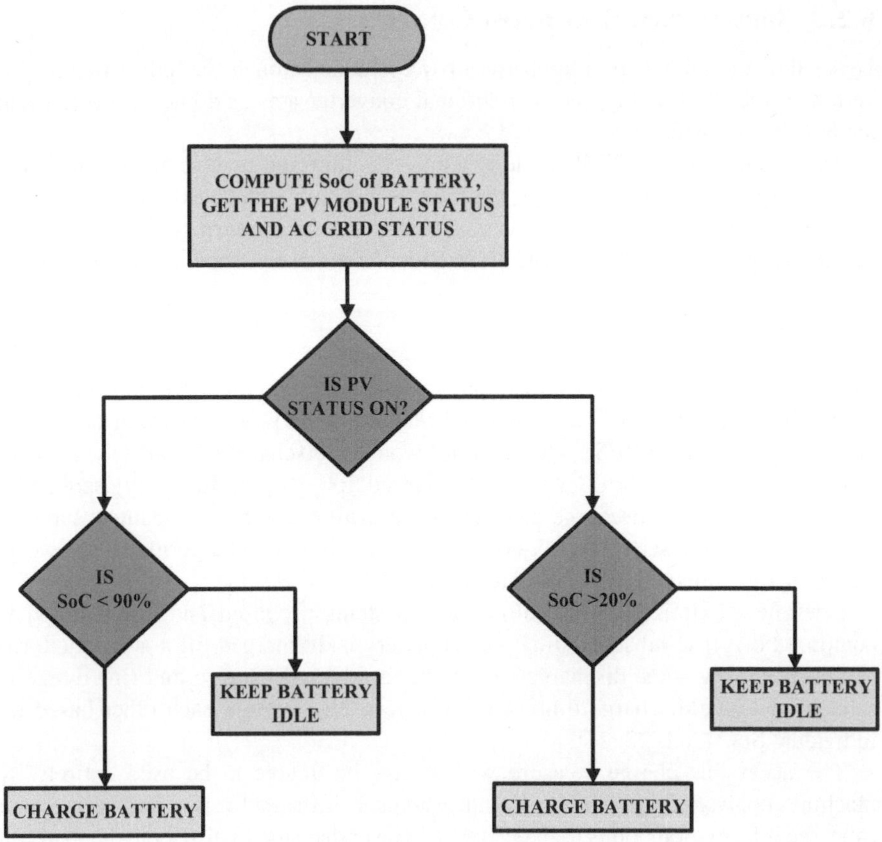

FIGURE 6.4 Flow chart representing the proposed control strategy.

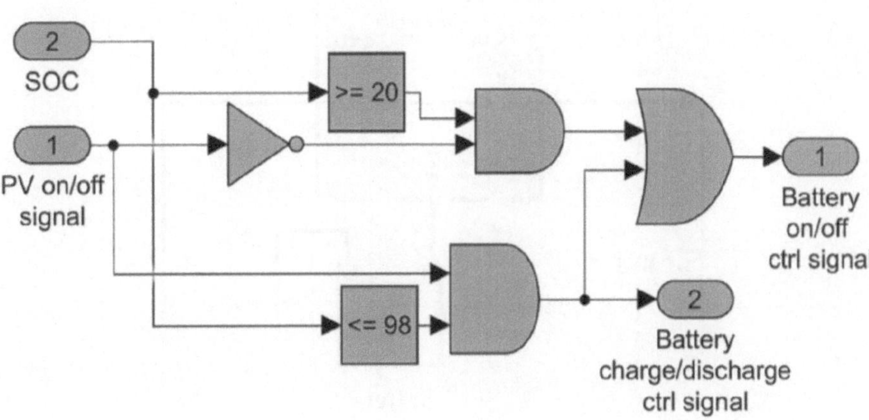

FIGURE 6.5 Implemented control algorithm for BESS.

6.2.2 Bidirectional Buck-Boost Converter

Power flow in and out from the battery bank is done through the bidirectional converter. While discharging, the bidirectional converter acts as a buck converter and feeds 48 V DC to the bus.

The discharging of BESS is done using the constant current mode. Similarly, when charging, the battery bank is charged by constant voltage till the SoC reaches 80%, and after this point, it will be charged by a constant current [18,19]. A traditional circuit diagram of a DC/DC bidirectional converter is shown in Figure 6.6.

6.2.3 BESS Scheduling

The smart distribution of each load demand over individual battery cells enhances battery lifetime. Power flow from and to the battery is via power electronic converter-based switches. In the BESS, all cell series won't be discharged completely; instead, those cells with higher SoC levels are discharged first. As proposed in [9], sequential and, in [10], parallel discharge algorithms are available. Only one battery supplies the load demand in sequential algorithms, whereas in parallel algorithms, the workload is shared among a subset of batteries [20,21].

Lifetime maximization of multi-battery systems proposed and tested in [9] is composed of: (i) serial scheduling – each battery is discharged till it is emptied; (ii) static scheduling – the discharge time is fixed for each battery; and (iii) dynamic scheduling – the discharge time of each battery differs from each other based on individual SoC level [22–24].

The accessible charge measure will always be unable to be used entirely by machines controlled by split battery subsystems as if the cell was solid. At the same time, the information points to the fact that a clever decision to distribute the load can help to reduce the efficiency hole between the solid and distributed strength supplies [15]. Because of the multi-battery framework, we have difficulty choosing which

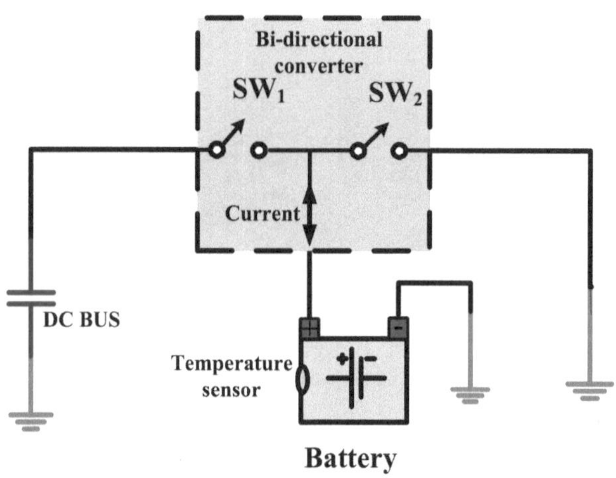

FIGURE 6.6 BESS-fed bidirectional buck-boost converter.

battery pack is required to combine to make the overall lifetime progress toward the best case, anywhere during the framework activity (i.e., solid battery). The battery booking question will be examined as the exhibition of the reservation process can be significantly affected by factors such as arrangement of the subsystem of battery (i.e., number and limit of battery packs in the supply of power) and role of the current heap (i.e., responsibility).

Series battery scheduling is of three types:

a. Serial Scheduling: The first type, known as sequential planning, releases each battery systematically. At the end of the day, a BESS is deactivated from the heap. This approach offers a single level of opportunity: the request to release the batteries. In case N packs are included in the battery subsystem, N! sequential bookings can be combined. Clearly, changing the battery requirement is essential only if there is no similar limit for any battery; responsibility for the long distance is not consistent.
b. Static Scheduling: This second type of arrangement, known as static planning, releases batteries under a cooperative plan. For a fixed time measurement, every battery remains associated with the heap and is then separated. It will be used again after the release of any remaining batteries for its relegated timeframe. Therefore, the policies of this class are described by two limits: the request for release of batteries and the time (or time-cutting) during which each battery is linked to the heap. Every pack may be awarded the amount of N! and MN's conceivable static bookings that should be examined if N battery packs and M time cuts are accepted. The minor difficult thing to do is to cut alone time on all batteries. Therefore, some pursuit should be taken out. The number of reservations to be examined decreases to N!
c. Dynamic Scheduling: The policies described use individual cell information (battery) to prioritize cell selection. These data include the physical state of individual batteries, expressed in terms of battery banking terminal voltage, charge status (SoC), and download time. The selected representative quantity should be observed on the battery interface in order to implement the proposed scheme in real time. These programs are classified as dynamic planning. There is no dynamic scheduling that is the time frame of each battery in the battery bank.

Instead, each battery has a different download time depending on the quantity observed. Dynamic scheduling is observed in this system and offers a fine-tuning of time slices, which rest and recover from the battery bank [25]. Dynamic programmatic arrangements are advantageous to adjust the idle time required for continuous discharge battery recovery. Dynamic scheduling is more effective in case of battery systems with different capacities. The possible combinations will be $2N-1$ for a total of N batteries in the battery bank.

The implementation of the dynamic scheduling for the proposed model considers a lower cutoff limit of 20%. This model consists of a battery bank with two separate batteries (cells) connected in series. The two batteries have the same parameters. When any battery enters the depth of discharge region and the battery is cut out of the grid, it does not enter the depth of release and the design control strategy is enabled.

If one battery, i.e., the SoC level of Battery-A or Battery-B, falls under 20%, the separation of each battery from the bus and typical operating performance of the rest of the system. The total charge on a DC bus is met with the leftover batteries when the battery is isolated [26]. The described circuit design is shown in the shape and dynamic scheduling of the system with suitable controlling elements shown in Figures 6.7 and 6.8.

6.3 DC LOADS

A DC home with DC appliances such as LED bulbs for lighting needs, fans with brushless DC motors, and cooking appliances such as an induction stove, cooler, and mini-refrigerator work on DC [27].

The wattage of all the connected loads to the DC bus of the proposed DC home is represented in Table 6.2. The total load, including all the appliances connected to the bus, is 2 kW. The proposed model is simulated for different loads, i.e., resistive and dynamic loads, as shown in Figure 6.9, with different switching combinations. The obtained results with appropriate analysis are discussed further.

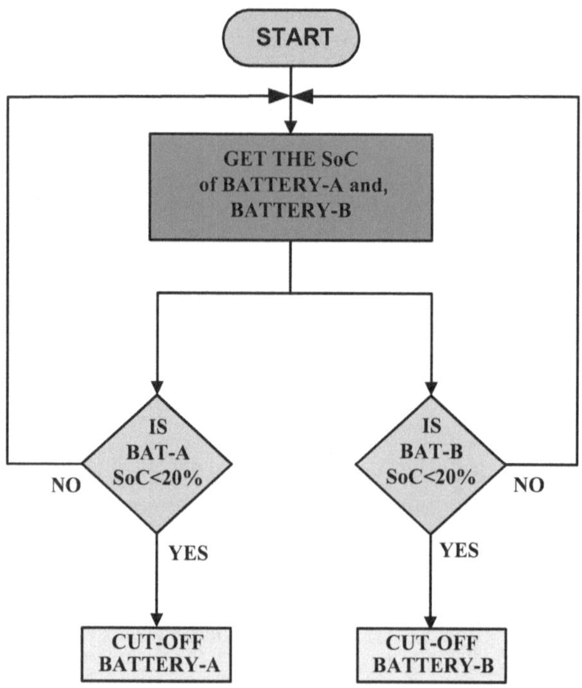

FIGURE 6.7 Flow chart of dynamic scheduling working.

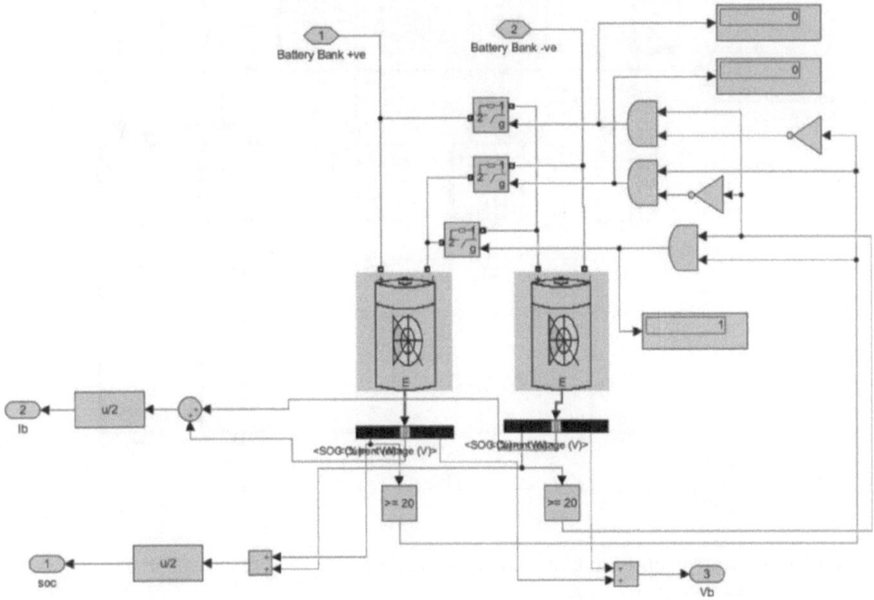

FIGURE 6.8 The proposed model.

TABLE 6.2
DC Load Specifications

Household Appliances	Quantity	Power Rating (Wh)
Fan	2	25
Light	5	5
Mobile phone	2	5
TV	1	25
Induction stove	1	400
Mixer	1	100
Total (including the number of hours of usage)		2 kW

6.4 CONTROL SCHEME OF POWER ELECTRONIC CONVERTERS

6.4.1 MPPT Buck Converter Control Scheme

The buck converter employed to achieve the required DC bus voltage of 48 V is shown in Figure 6.10. Whenever the PV is ON, i.e., during the daytime, the DC bus is fed from the PV array. So, there is a need to step down the PV output voltage to the required DC bus voltage level.

The buck converter shown in Figure 6.10 is designed using an IGBT. The diode is used for the freewheeling mode of operation. The switching of a semiconductor is decided by the closed-loop PWM controller shown in Figure 6.11. The values of the controller parameters are shown in Table 6.3.

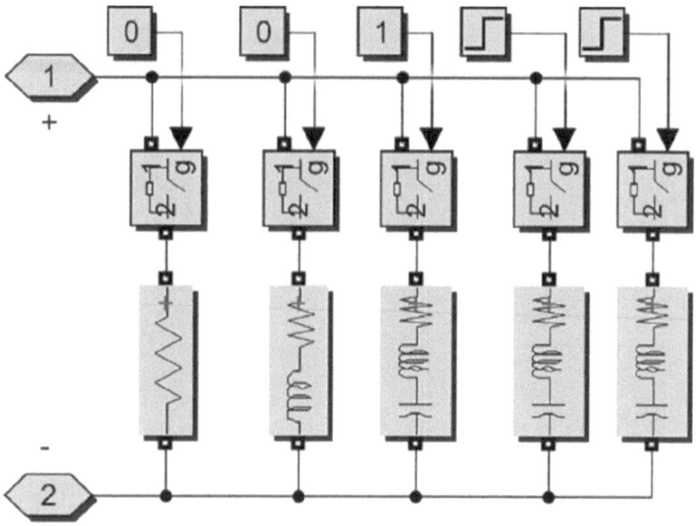

FIGURE 6.9 Implemented dynamic load for the proposed DC home structure.

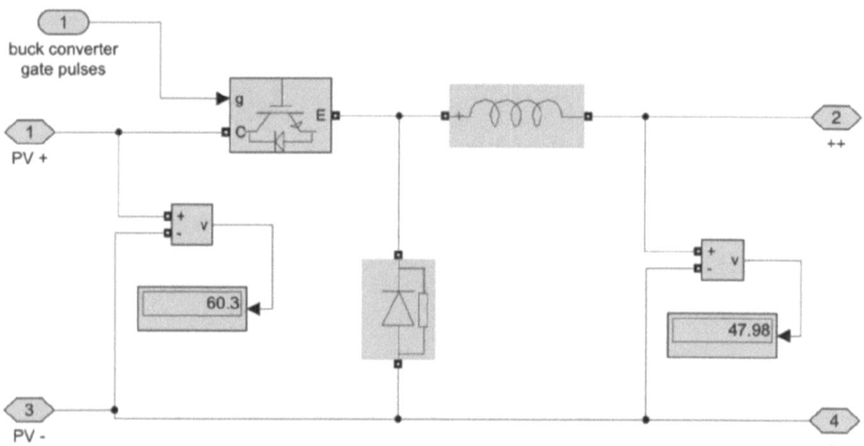

FIGURE 6.10 Buck converter for MPPT algorithm application.

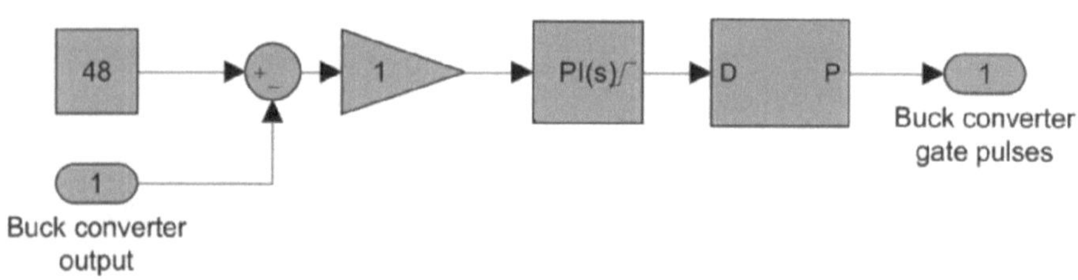

FIGURE 6.11 Buck converter closed-loop PWM controller.

TABLE 6.3
MPPT Charge Controller Parameters

Controller Parameter	Parameter Value
K_P	0.15
K_i	50
K_d	0
Gain	1

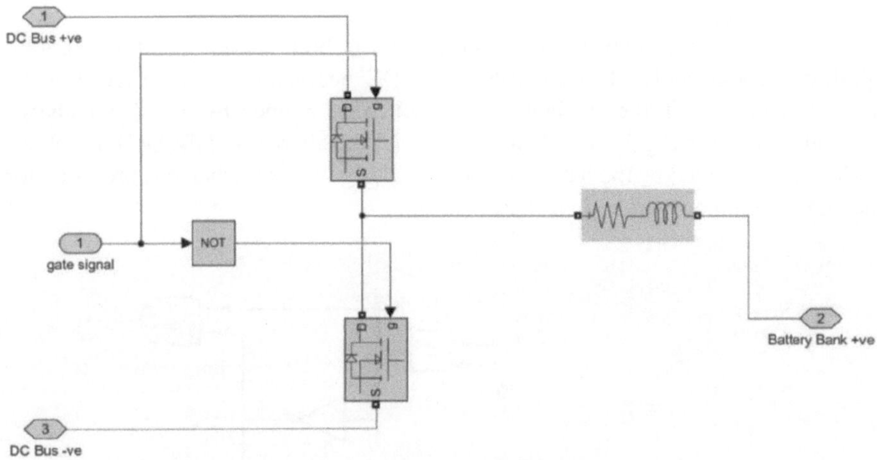

FIGURE 6.12 Implemented bidirectional buck-boost converter for the proposed model.

6.4.2 Implementation of Bidirectional Converter

A two-switch bidirectional converter [12] is implemented to control the in and out power flow through the battery bank during charging and discharging applications. The bidirectional converter implemented in the developed model is shown in Figure 6.12.

6.4.3 Control Strategy of the Bidirectional Converter during Charging/Discharging of Battery Pack

The internal current control loop and external voltage control loop are the two loops proposed in the implemented scheme. The inner current loop mitigates fluctuations in battery current due to parameter variations, and the profile of DC bus voltage is delimited by the outer voltage loop [12,14].

If the DC-link tension (V_{dc}) is higher than the V_{dc}, S_1 is turned ON as a buck converter, and when the DC-link voltage (V_{dc}) is higher than the reference-point-reference tension (V_{dcref}), $S2$ is switched ON as the boost converter to operate the system (discharging mode). Therefore, the S_1 and S_2 switches are alternately activated to achieve or improve the necessary mode of operation [15,19,28,29]. The strategy to control the DC-to-DC converter in two directions is shown in Figure 6.12. Compared to set point or DC bus voltage (V_{dcref}), the converter output voltage (V_{dc}) gives the error signal that is passed through the PI controller to generate reference battery current (I_{batref}). This I_{batref} current is then compared to the sensed battery current (I_{bat}) to get the error signal. A PI controller is set to obtain a control signal, as shown in Figure 6.12. The contrast between the control signal received and a three-wheel high-frequency signal provides the PWM pulse needed for gate control of S_1 and S_2 switches. The DC-to-DC converter's switching frequency is 20 kHz. The designed external voltage loop controller with the PID control for charging and discharging the battery bank is shown in Figures 6.13 and 6.14, respectively.

The values of the controller parameters used in both charging and discharging applications are tabulated in Table 6.4. The DC bus voltage is maintained at the desired voltage when the available PV output is maintained by the MPPT charge controller shown in Figure 6.10. When PV out is unavailable, and the DC bus voltage profile is maintained by the battery bank, a two-stage control scheme is proposed for the bidirectional converter control [30].

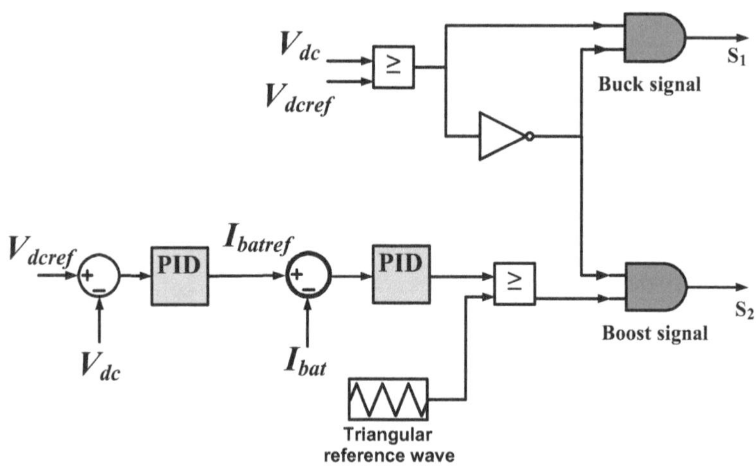

FIGURE 6.13 Control of a bidirectional buck-boost converter.

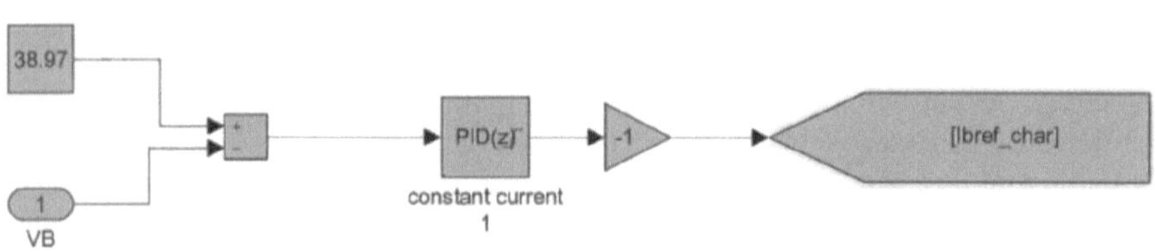

FIGURE 6.14 Battery bank charging control strategy.

TABLE 6.4
Controller Parameters for Bidirectional DC-to-DC Converter

Charging Controller		Discharging Controller	
Controller Parameter	**Parameter Value**	**Controller Parameter**	**Parameter Value**
K_p	40	K_p	0.25
K_i	2000	K_i	100
K_d	0	K_d	0
Gain	1	Gain	0.25

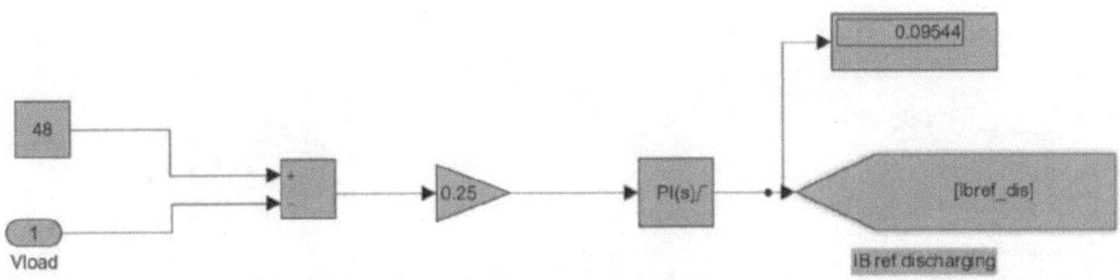

FIGURE 6.15 Battery bank discharging control strategy.

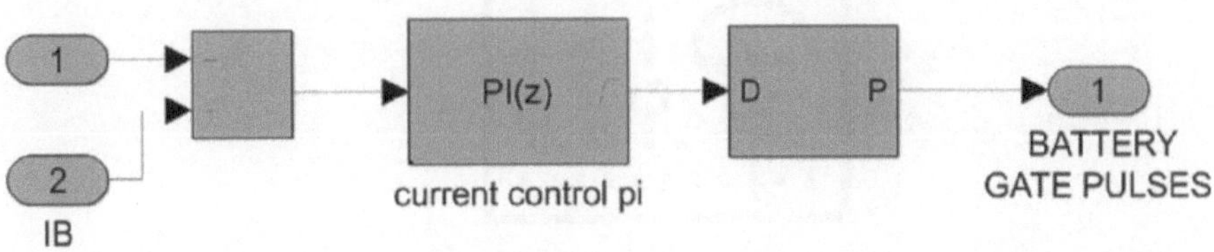

FIGURE 6.16 Inner current loop of bidirectional converter for DC bus voltage profile maintenance. The values of the parameters of the controller used are shown in Table 6.5.

6.4.4 DC Bus Voltage Regulation Scheme

The first is the outer voltage loop, which is different for charging and discharging applications. Then comes the inner current loop shown in Figure 6.16, which takes the voltage loop output as the reference battery current (I_{batref}) and compares it with the actual battery current (I_B) and gives this error signal as the input to the controller, and then the PWM pulses are generated based on the controller output. The switching frequency is chosen to be 10 kHz to get a smooth output voltage waveform from the battery. This control strategy implementation is shown in Figure 6.16.

6.5 SIMULATION RESULTS

The proposed BMS algorithm is implemented for the DC home configuration shown in Figure 6.4. A model is designed in MATLAB Simulink platform as in Figure 6.15

TABLE 6.5
Bidirectional Converter Outer Current Loop Controller Parameters

Controller Parameter	Chosen Value
K_P	0.25
K_i	100
K_d	0
Gain	1

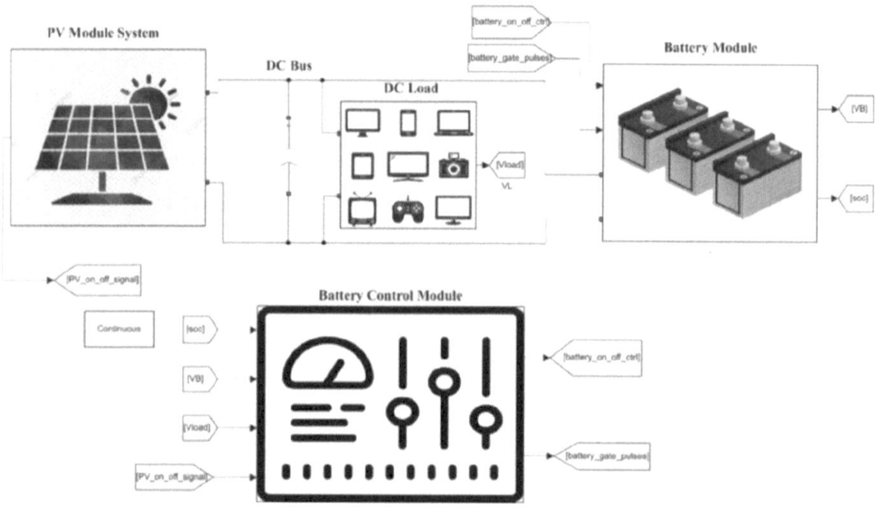

FIGURE 6.17 Implementation of the proposed control strategy in MATLAB.

for checking the performance of the developed control strategy. The developed system acts as a stand-alone system with two energy sources, i.e., rooftop solar PV installation and battery bank.

To know the performance of the proposed system in different load variations and energy source variations, eight different modes of operations are explained in detail below.

6.5.1 Mode 1: DC Bus Fed from PV Array

This mode is during the daytime. In this case, the PV will be functional and supply power to the DC bus by harvesting solar energy. The DC bus voltage obtained is 48 V_{DC}. The system voltage got stabilized to 48 V at 0.03 seconds with a tolerance of 2% (±0.5 V) as shown in Figure 6.16.

During the first mode, the SoC levels of individual batteries are chosen as 40% such that both will get charged as in Figure 6.17, and the overall battery bank characteristics, i.e., SoC, terminal voltage, and current in this mode, are shown in Figure 6.18.

6.5.2 Mode 2: DC Bus Fed from Battery Bank

This mode is during the nighttime or when the PV output is not available, for example during a rainy day. In this case, the battery bank will be functional and supply power to the DC bus. The DC bus voltage obtained is 48 V_{DC}. The system voltage got stabilized to 48 V at 0.3 seconds with a tolerance of 2% (±1 V), and finally, by 0.4 seconds, the value stabilized with a tolerance of 0.02% (±0.01 V) as shown in Figure 6.19.

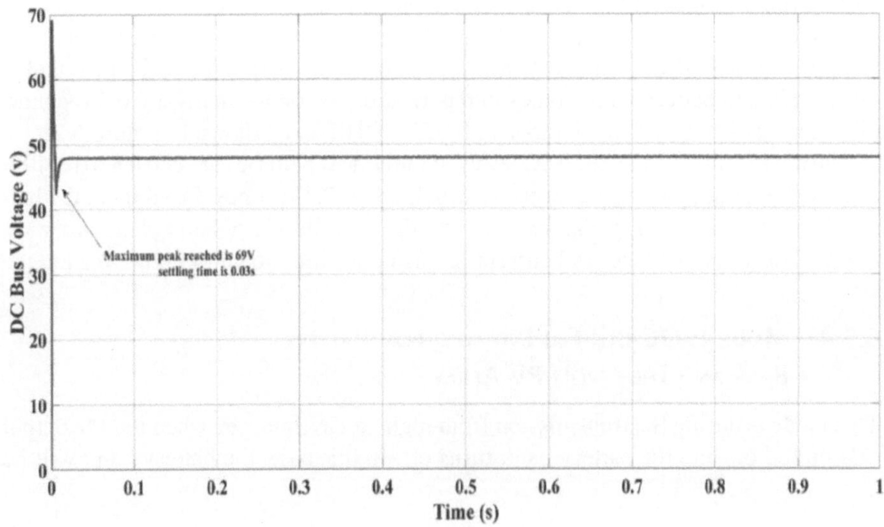

FIGURE 6.18 DC bus voltage while being fed from solar PV output during daytime.

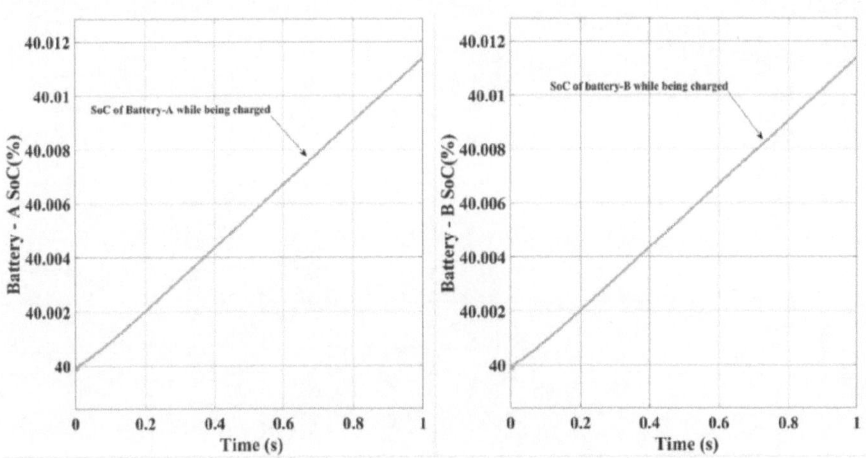

FIGURE 6.19 Individual battery SoC.

During the second mode, the SoC levels of individual batteries are chosen to be 40% each such that both will get charged, as in Figure 6.20. The overall battery bank characteristics, i.e., SoC, terminal voltage, and current in this mode, are shown in Figure 6.21.

6.5.3 Mode 3: DC Bus Fed Initially from PV and Then from Battery Bank

This mode is during the transmission from daytime to nighttime, i.e., when the PV output gets turned OFF and the battery bank turns ON. In this case, the battery bank will be functional from 0.3 seconds and supply power to the DC bus. The DC bus voltage obtained is 48 V_{DC}. The system voltage while being fed from PV got stabilized to 48 V at 0.043 seconds with a tolerance of 0.02% (±0.01 V), and when the PV turns OFF, and battery bank comes into play at 0.3 seconds; finally, by 0.5 seconds, the value stabilized with a tolerance of 0.02% (±0.01 V) as shown in Figure 6.22.

During the third mode, the SoC levels of individual batteries are chosen to be 40% each such that both will get charged during the first 0.3 seconds. From then, the battery bank discharges as in Figure 6.23, and the overall battery bank characteristics, i.e., SoC, terminal voltage, and current in this mode, are shown in Figure 6.24.

6.5.4 Mode 4: DC Bus Fed Initially from Battery Bank and Then from PV Array

This mode is during the transmission from night to daytime, i.e., when the PV output gets turned on, and the battery bank turns off. In this case, the battery bank will be

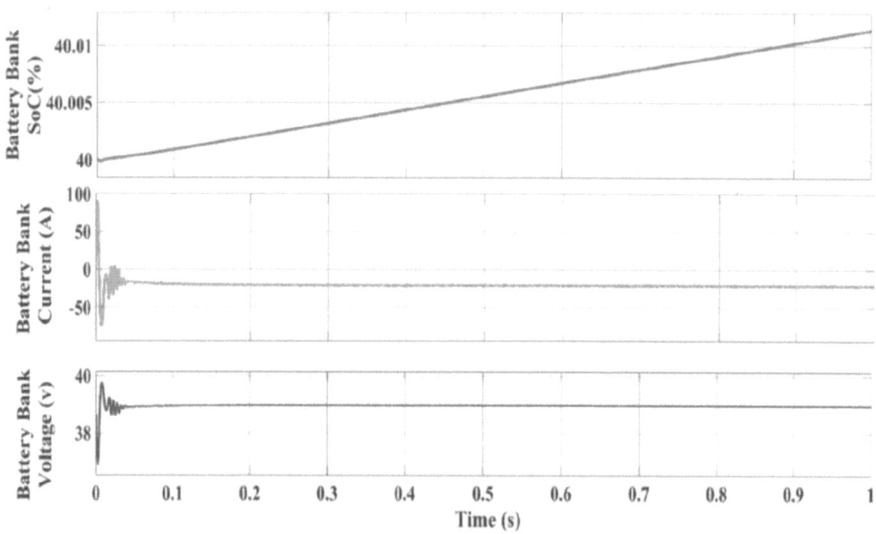

FIGURE 6.20 Battery bank SoC, terminal voltage, and current while being charged from PV through DC bus.

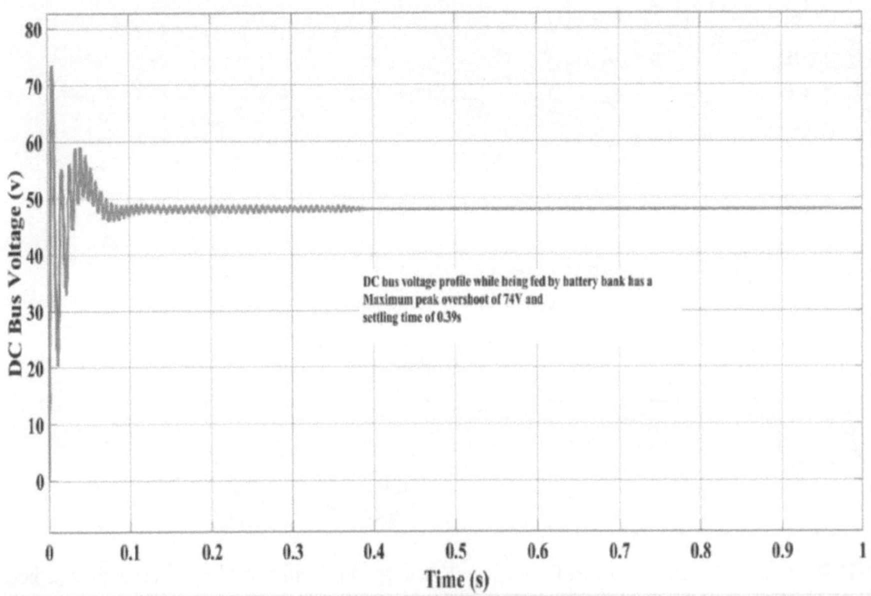

FIGURE 6.21 DC bus voltage while being fed from battery bank.

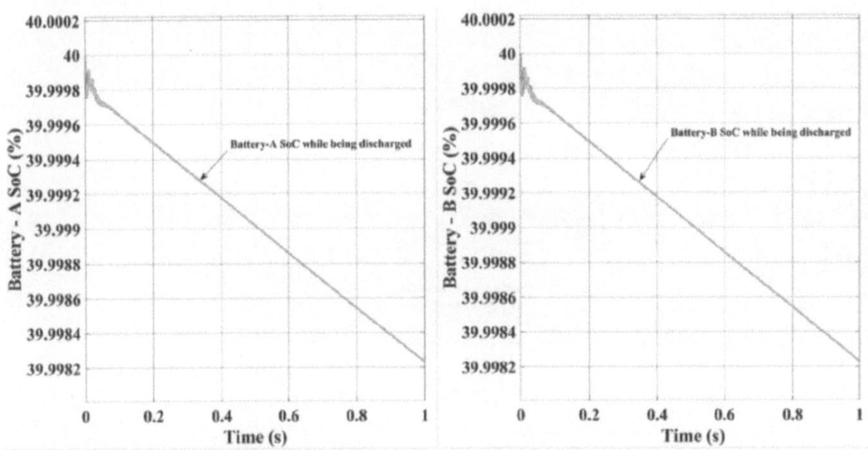

FIGURE 6.22 Individual battery SoC.

functional till 0.6 seconds and supply power to the DC bus, and from then, PV feeds power to the DC bus by harvesting solar energy. The DC bus voltage obtained is 48 V_{DC}. The system voltage while being fed from battery bank got stabilized to 48 V at 0.38 seconds with a tolerance of 0.02% (±0.01 V), and when the PV turns ON and the battery bank turns OFF at 0.6 seconds, finally by 0.62 seconds, the value stabilized with a tolerance of 0.02% (±0.01 V) as shown in Figure 6.25.

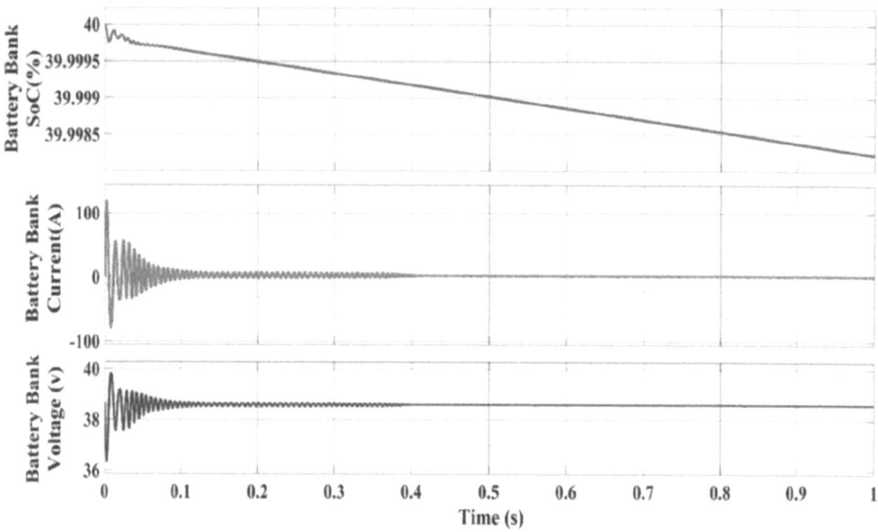

FIGURE 6.23 Battery bank SoC, terminal voltage, and current while being discharged through DC bus.

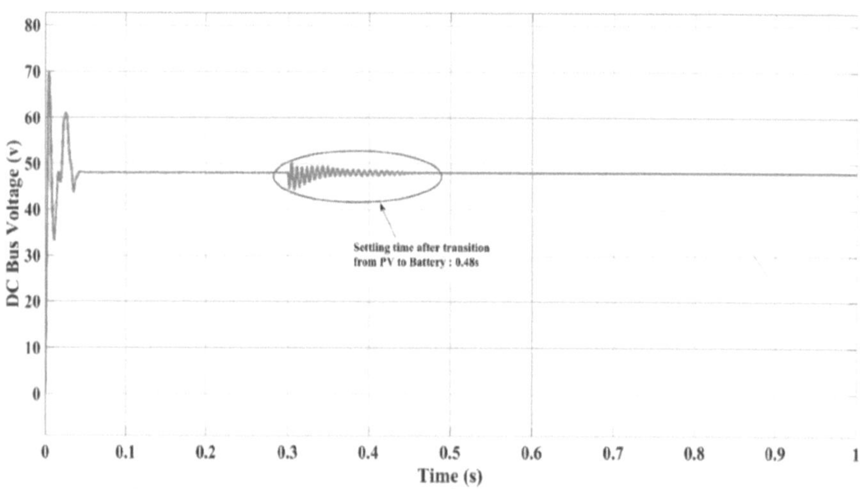

FIGURE 6.24 DC bus voltage while initially being fed from PV till 0.3 seconds and then by battery bank.

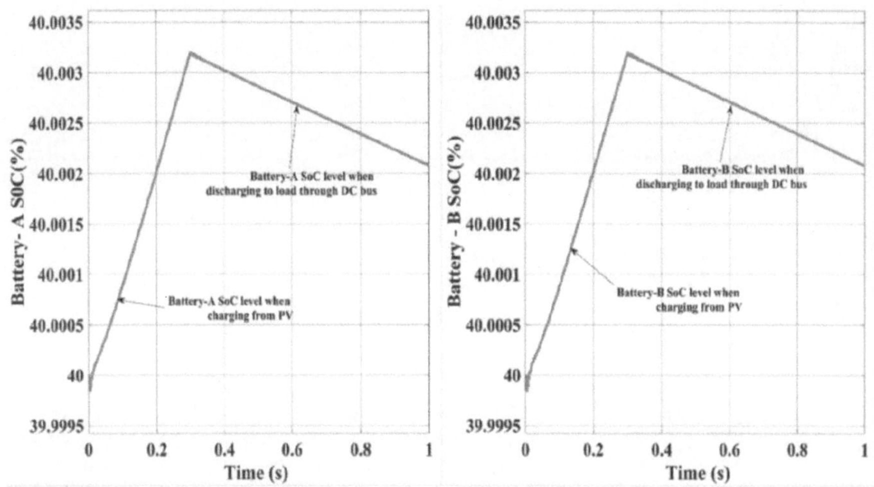

FIGURE 6.25 Individual battery SoC.

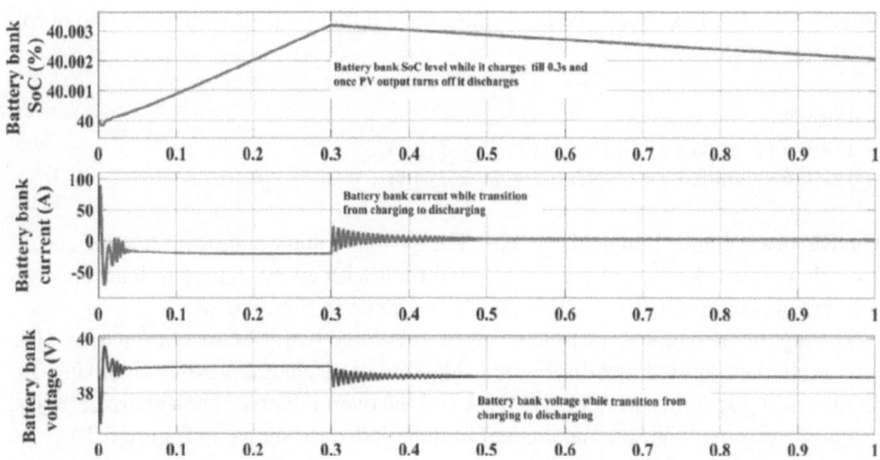

FIGURE 6.26 Battery bank SoC, terminal voltage, and current are initially being charged and then discharged through the DC bus.

During the fourth mode, the SoC levels of individual batteries are chosen to be 40% each such that both will get discharged during the first 0.6 seconds. From then, the battery bank charges as in Figure 6.26, and the overall battery bank characteristics, i.e., SoC, terminal voltage, and current in this mode, are shown in Figure 6.27.

6.5.5 Mode 5: DC Bus Fed Initially from Battery Bank with Dynamic Load

This mode is during the nighttime or when the PV output is not available, for example during a rainy day. In this case, the battery bank will be functional and supply

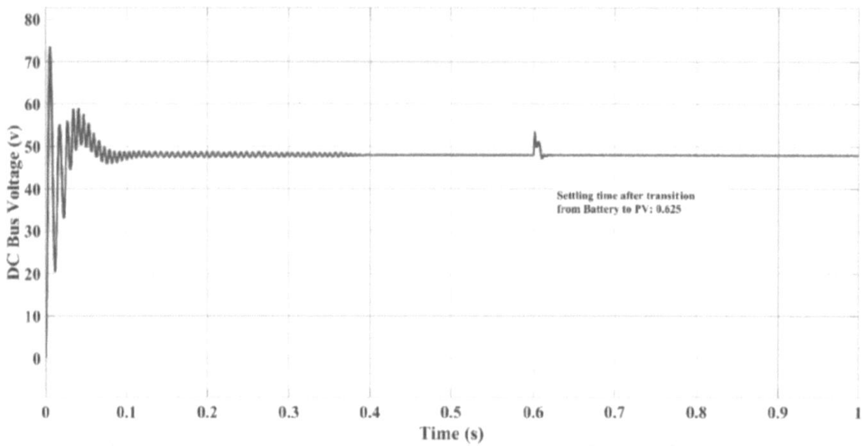

FIGURE 6.27 DC bus voltage while initially being fed from PV till 0.6 seconds and then by battery bank.

power to the DC bus. Here, the load chosen is a dynamic load with switching at following instances:

For $0 < t < 0.3$ seconds, $P_L = 150$ W.
For 0.3 seconds $< t < 0.6$ seconds, $P_L = 270$ W.
For 0.6 seconds $< t < 0.9$ seconds, $P_L = 750$ W.

The DC bus voltage obtained is 48 V_{DC}. The system voltage stabilized to 48 V at 0.27 seconds with a tolerance of 2% (±1 V), and then with every change in load, the system regains stability within 0.03 seconds, as shown in Figure 6.28.

During the fifth mode, the SoC levels of individual batteries are chosen to be 40% each such that both will get discharged. The rate of discharge is different during different loading sections, as in Figure 6.29, and the overall battery bank characteristics, i.e., SoC, terminal voltage, and current in this mode, are shown in Figure 6.30.

6.5.6 Mode 6: DC Bus Fed from PV Array with Dynamic Load

This mode is during the daytime or when the PV is ON. In this case, the battery bank will harvest power from solar energy and supply power to the DC bus. Here, the load chosen is a dynamic load with switching at following instances:

For $0 < t < 0.3$, $P_L = 150$ W.
For $0.3 < t < 0.6$, $P_L = 270$ W.
For $0 < t < 0.3$, $P_L = 750$ W.

The DC bus voltage obtained is 48 V_{DC}. The system voltage stabilized to 48 V at 0.1 seconds with a tolerance of 2% (±1 V), and then with every change in the load, the system regains stability within 0.01 seconds as shown in Figure 6.31.

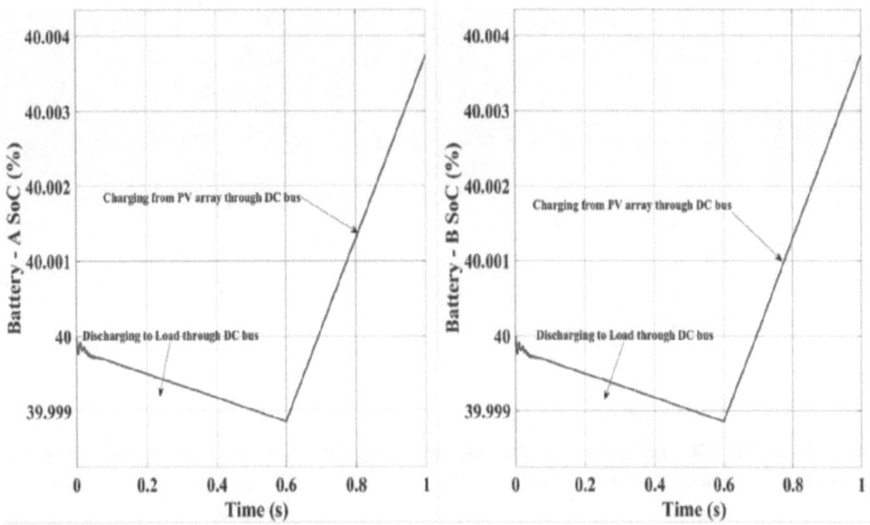

FIGURE 6.28 Individual battery SoC.

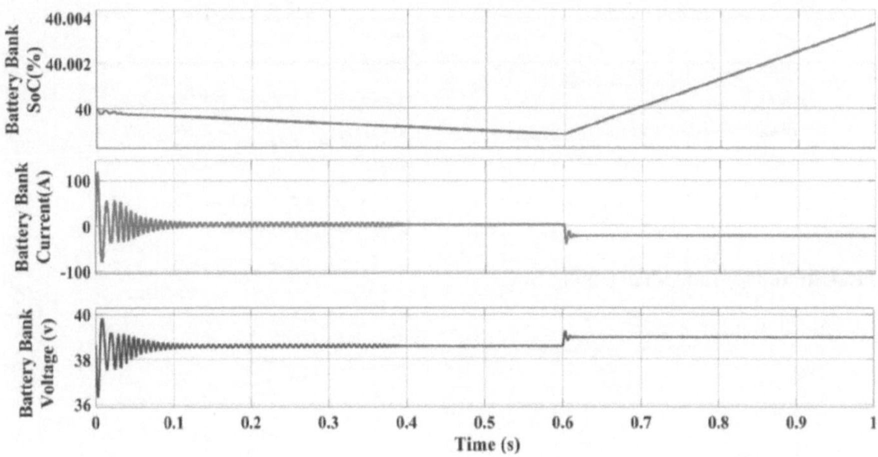

FIGURE 6.29 Battery bank SoC, terminal voltage, and current are initially discharged and then charged through DC bus.

During the sixth mode, the SoC levels of individual batteries are chosen to be 40% each such that both will get charged (Figure 6.32). The rate of charge is the same during different loading sections, as in Figure 6.33, because load shifting does not affect the battery bank since the battery bank is also a load during the charging case. The overall battery bank characteristics, i.e., SoC, terminal voltage, and current in this mode, are shown in Figure 6.34.

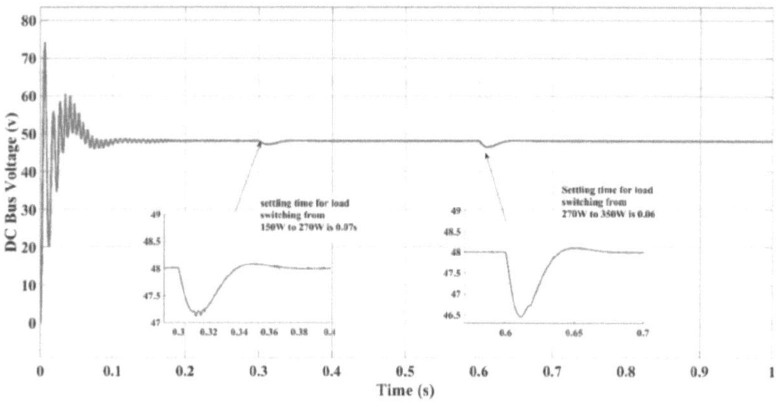

FIGURE 6.30 DC bus voltage while initially being fed from the battery bank to supply power to the connected dynamic load.

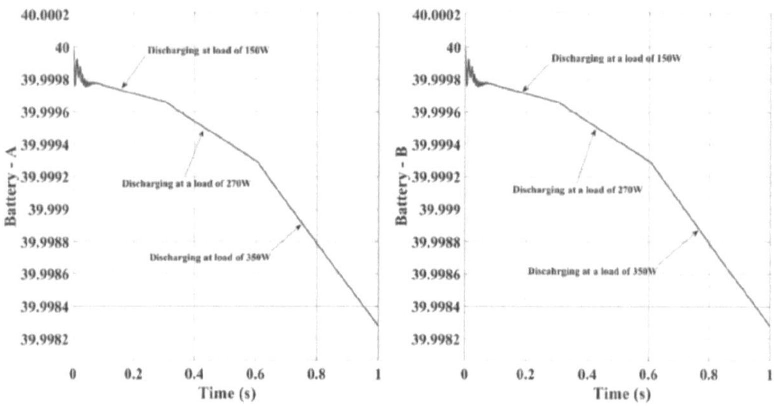

FIGURE 6.31 Individual battery SoC.

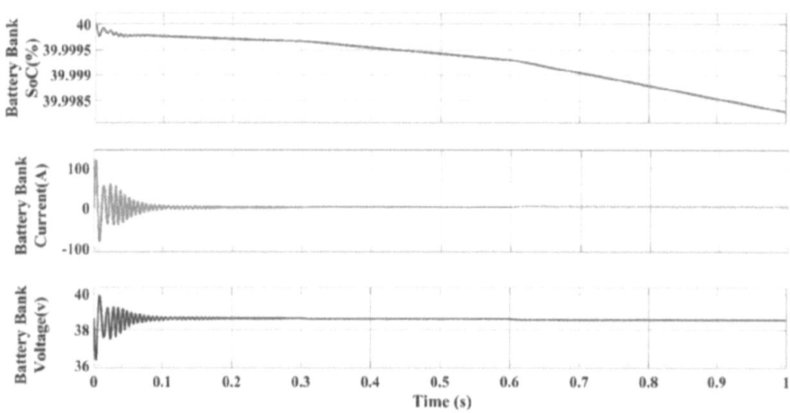

FIGURE 6.32 Battery bank SoC, terminal voltage, and current while being discharged through DC bus to the designed dynamic load.

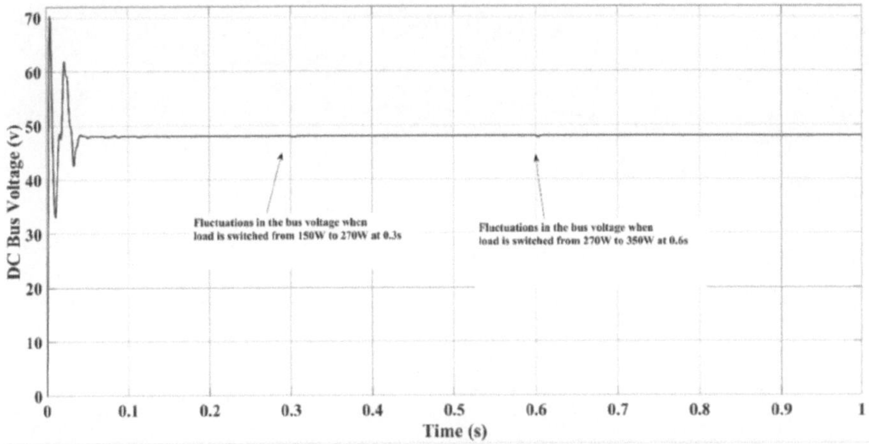

FIGURE 6.33 DC bus voltage while initially being fed from PV array to supply power to the connected dynamic load.

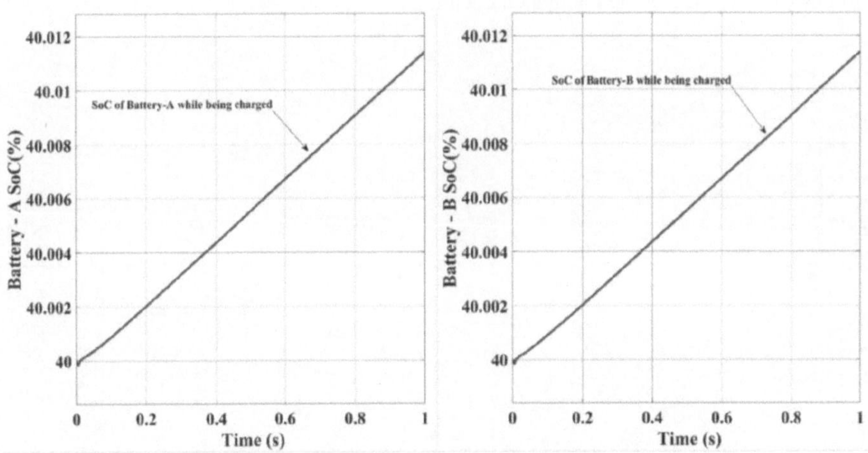

FIGURE 6.34 Individual battery SoC.

6.5.7 Mode 7: DC Bus Fed Initially from Battery Bank with Battery Scheduling

This mode is considered during the nighttime; i.e., the DC bus is being fed power by the battery bank. Here, to trigger the proposed battery scheduling algorithm, the SoC of one of the batteries (Battery-B) is set at 20.005%. The DC bus voltage during this mode is shown in Figure 6.35. Once Battery-B cuts off from the system at 0.5 seconds, the total load demand is met by Battery-A itself, and hence, a voltage spike occurs at that time. The system gets stabilized within 0.03 seconds after the battery scheduling algorithm kicks in.

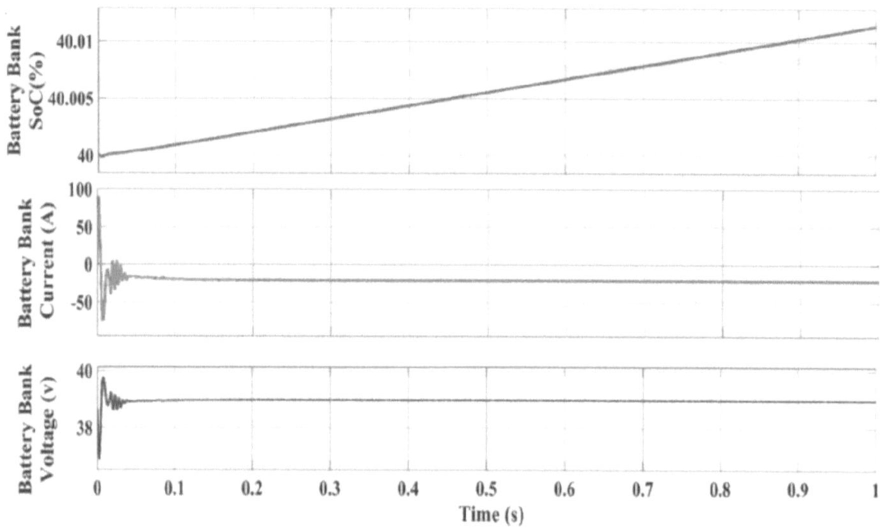

FIGURE 6.35 Battery bank SoC, terminal voltage, and current while being charged through DC bus while PV is feeding the dynamic load.

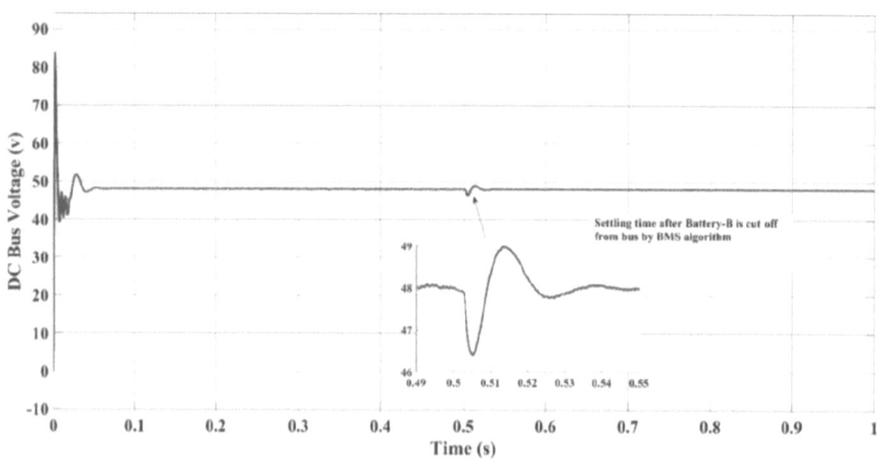

FIGURE 6.36 DC bus voltage being fed from battery bank with battery scheduling algorithm implemented.

During the seventh mode, the SoC levels of individual batteries are chosen to be 40% for Battery-A and 20.0025 for Battery-B such that both will get discharged (Figure 6.36). The rate of discharge is increased once Battery-B cuts off from the system, as in Figure 6.37, because the total load demand is met by Battery-A, and the overall battery bank characteristics, i.e., SoC, terminal voltage, and current in this mode, are shown in Figure 6.38.

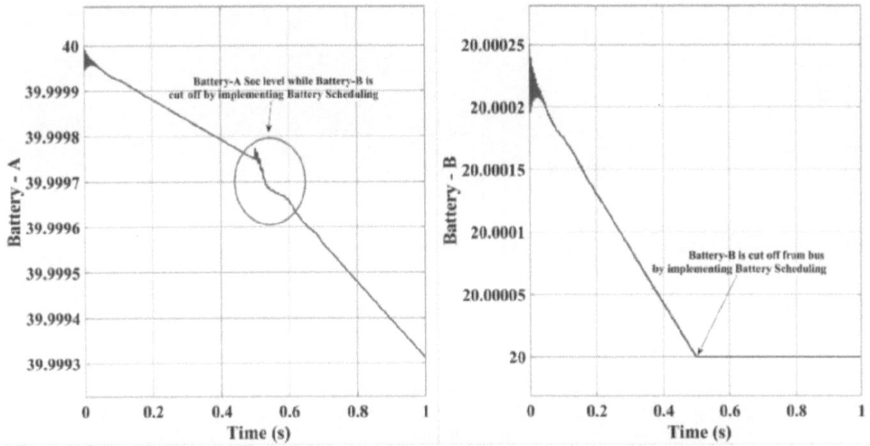

FIGURE 6.37 Individual through DC bus with battery scheduling algorithm applied.

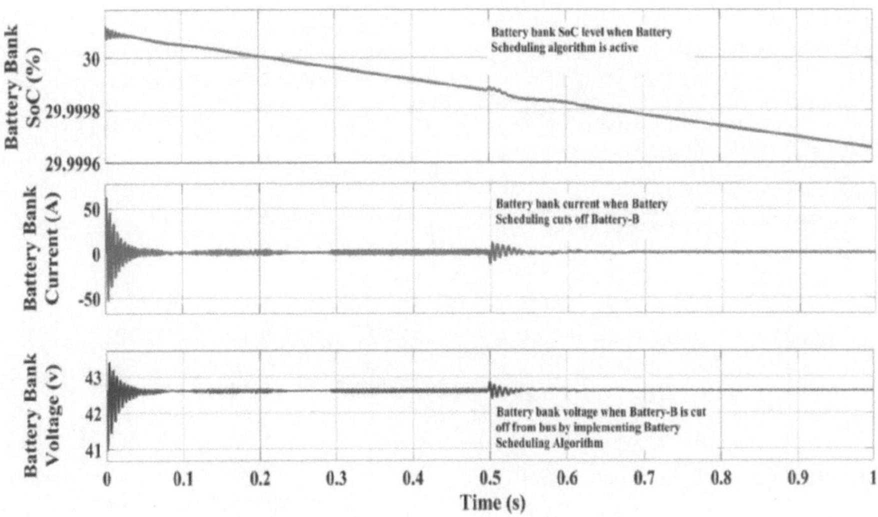

FIGURE 6.38 Battery bank SoC, terminal voltage, and current while being charged through DC bus with battery management scheduling applied.

6.5.8 Mode 8: DC Bus Fed Initially from Battery Bank with Battery Management

This mode is considered during the nighttime; i.e., the DC bus is being fed power by the battery bank. Here, to trigger the proposed battery management algorithm, the SoC of the batteries (both Battery-A and Battery-B) is around 89.995% (Figure 6.39).

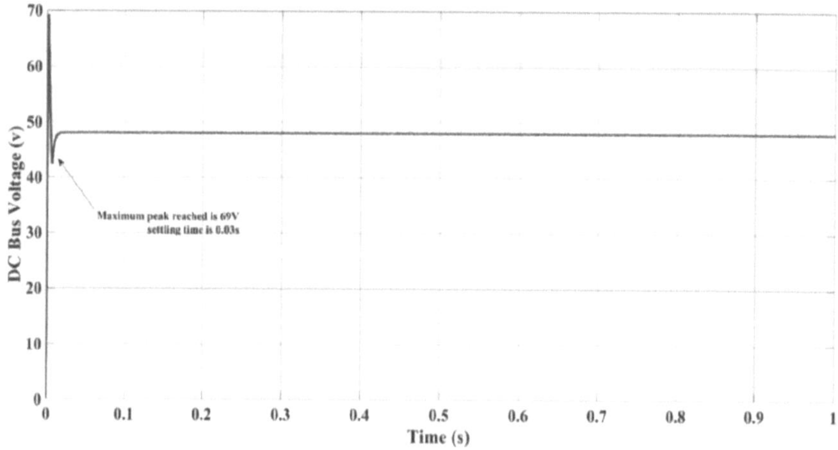

FIGURE 6.39 DC bus voltage being fed from PV with battery management algorithm implemented.

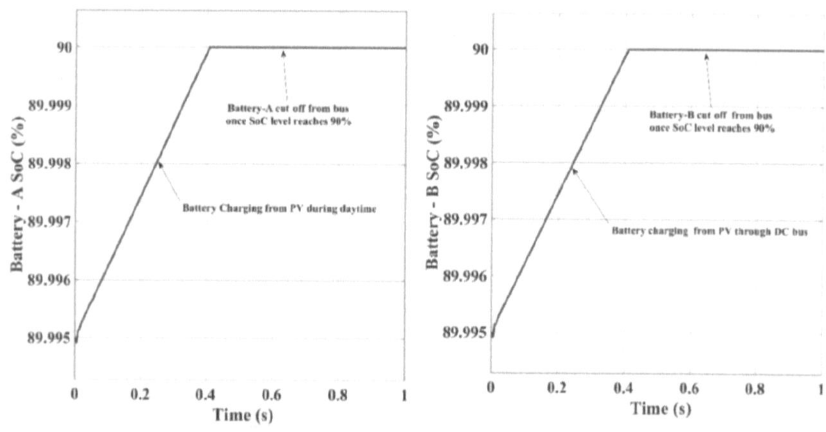

FIGURE 6.40 Individual battery SoC.

During the eighth mode, the SoC levels of individual batteries are chosen to be 89.995% each such that both will get charged. The rate of charge is the same till their SoC level reaches 90%, which is the upper cutoff limit set in the battery management algorithm and cuts off the battery bank from the bus as in Figure 6.40. The overall battery bank characteristics, i.e., SoC, terminal voltage, and current in this mode, are shown in Figure 6.41.

6.6 CONCLUSIONS

The proposed BMS algorithms are verified by varying the sources between PV array and battery bank. For analyzing the DC bus voltage profile, switching from PV array

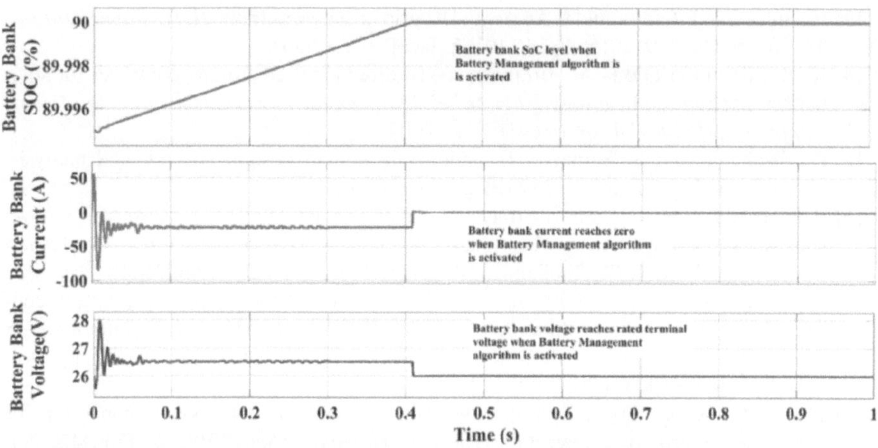

FIGURE 6.41 Individual battery SoC and battery bank SoC, terminal voltage, and current while being charged through DC bus with battery management algorithm applied.

input to battery bank and battery bank to PV array is performed virtually and the results provided in modes 3 and 4, respectively, are obtained. The obtained outputs prove that the system reaches a steady state within 0.03 seconds. Similarly, for PV array input, the DC bus performance is analyzed for a simple RL load. For a varying or dynamic load, which combines both R and RL loads as discussed in previous sections, the individual results are provided in modes 1 and 6. The obtained results prove the efficiency of the developed PV charge control algorithm in regulating the DC bus voltage while being fed from PV. Modes 2 and 5 discussed the DC bus voltage profile stability while being supplied from the battery bank for both R and dynamic loads. The main aim of a BMS is to continuously monitor and maintain individual battery health in the battery bank, which is achieved perfectly by the proposed BMS as shown in modes 7 and 8.

According to the findings, the suggested BMS for the PV system gives effective voltage regulation and exhibits strong dynamic stability during various operating modes and switching surges in the voltage. With a maximum peak overshoot of 9.25% at a peak time (T_P) of 4.7 ms, the proposed model displays an excellent dynamic response. The achieved settling time (T_S) is 0.03 seconds. As a result of the proposed BMS implementation, the battery lifetime can increase to 15.58% of the expected tenure.

REFERENCES

[1] A. Jhunjhunwala, A. Lolla, and P. Kaur, Solar-DC microgrid for Indian homes: A transforming power scenario, *IEEE Electrif. Mag.*, 4, 2, pp. 10–19, 2016, doi: 10.1109/MELE.2016.2543950.

[2] A. Jhunjhunwala and P. Kaur, Solar energy, dc distribution, and microgrids: Ensuring quality power in Rural India, *IEEE Electrif. Mag.*, 6, 4, pp. 32–39, 2018, doi: 10.1109/MELE.2018.2871277.

[3] P. Sanjeev, N. P. Padhy, and P. Agarwal, DC grid initiative in India, *IFAC-PapersOnLine*, 48, 30, pp. 114–119, 2015, doi: 10.1016/j.ifacol.2015.12.363.

[4] K. H. Edelmoser and F. A. Himmelstoss, Bi-directional DC-to-DC converter for solar battery baakup applications, *2004 IEEE 35th Annual Power Electronics Specialists Conference (IEEE Cat. No.04CH37551)*, 2004.

[5] A. El-Shahat and S. Sumaiya, DC-microgrid system design, control, and analysis, *Electron.*, 8, 2, 2019, doi: 10.3390/electronics8020124.

[6] Z. Liao and X. Ruan, Control strategy of bi-directional DC/DC converter for a novel stand-alone photovoltaic power system, *2008 IEEE Vehicle Power and Propulsion Conference VPPC 2008*, 2008, pp. 3–8, doi: 10.1109/VPPC.2008.4677404.

[7] P. Shaw, P. K. Sahu, S. Maity, and P. Kumar, Modeling and control of a battery connected standalone photovoltaic system, *2016 IEEE 1st International Conference on Power Electronics, Intelligent Control and Energy Systems (ICPEICES)*, 2016, pp. 1–6, doi: 10.1109/ICPEICES.2016.7853123.

[8] Sladić, S., Kolić, D., and Šuljić, M. Bidirectional DC/DC power converter for hybrid yacht propulsion system. *J. Marit. Transport. Sci.* 2018, 2, 133–142. doi: 10.18048/2018-00.133.

[9] L. Benini, G. Castelli, A. Macii, E. Macii, M. Poncino, and R. Scarsi, Extending lifetime of portable systems by battery scheduling, *Proceedings Design, Automation and Test in Europe. Conference and Exhibition 2001*, 2001, pp. 197–201, doi: 10.1109/DATE.2001.915024.

[10] R. Adany, D. Aurbach, and S. Kraus, Switching algorithms for extending battery life in electric vehicles, *J. Power Sources*, 2013, 231, pp. 50–59, doi: 10.1016/j.jpowsour.2012.12.075.

[11] C. Wang, W. Chen, S. Shao, Z. Chen, B. Zhu, and H. Li, Energy management of stand-alone hybrid PV system, *Energy Proc.*, 2011, 12, pp. 471–479, doi: 10.1016/j.egypro.2011.10.063.

[12] C. Kalavalli, K. Parkavikathirvelu, and R. Balasubramanian, Single phase bidirectional PWM converter for microgrid system. *Int. J. Eng. Technol.*, 2013, 5, 3, pp. 2436–2441.

[13] X. Qian, Y. Yang, C. Li, and S. C. Tan, Operating cost reduction of DC microgrids under real-time pricing using adaptive differential evolution algorithm, *IEEE Access*, 8, pp. 169247–169258, 2020, doi: 10.1109/ACCESS.2020.3024112.

[14] Y.-K. Chen, Y.-C. Wu, C.-C. Song, and Y.-S. Chen, Design and implementation of energy management system with fuzzy control for DC micro-grid systems, *IEEE Trans. Power Electron.*, 28, 4, pp. 1563–1570, 2011.

[15] N. M. L. Tan, T. Abe, and H. Akagi, Design and performance of a bidirectional isolated DC-DC converter for a battery energy storage system, *IEEE Trans. Power Electron.*, 27, 3, pp. 1237–1248, 2012, doi: 10.1109/TPEL.2011.2108317.

[16] H. S. Kim, M. H. Ryu, J. W. Baek, and J. H. Jung, High-efficiency isolated bidirectional AC-DC converter for a DC distribution system, *IEEE Trans. Power Electron.*, 28, 4, pp. 1642–1654, 2013, doi: 10.1109/TPEL.2012.2213347.

[17] S. Sinha and P. Bajpai, Power management of hybrid energy storage system in a standalone DC microgrid, *J. Energy Storage*, 30, 2020, doi: 10.1016/j.est.2020.101523.

[18] A. Mohamed, M. Elshaer, and O. Mohammed, Bi-directional AC-DC/DC-AC converter for power sharing of hybrid AC/DC systems, *IEEE Power & Energy Society General Meeting*, pp. 1–8, 2011, doi: 10.1109/PES.2011.6039868.

[19] L. Zhang, T. Wu, Y. Xing, K. Sun, and J. M. Gurrero, Power control of DC microgrid using DC bus signaling, *2019 IEEE Applied Power Electronics Conference and Exposition (APEC)*, pp. 1926–1932, 2011, doi: 10.1109/APEC.2011.5744859.

[20] S. Duryea, S. Islam, and W. Lawrance, A battery management system for standalone photovoltaic energy systems, *IEEE Ind. Appl. Mag.*, 7, 3, pp. 67–72, 2001, doi: 10.1109/2943.922452.

[21] S. Sinha, A. K. Sinha, and P. Bajpai, Solar PV fed standalone DC microgrid with hybrid energy storage system, *2017 6th International Conference on Computer Applications in Electrical Engineering-Recent Advances (CERA)*, 2018, pp. 31–36, 2018, doi: 10.1109/CERA.2017.8343296.

[22] A. Sanal, V. Mohan, M. R. Sindhu and S. K. Kottayil, Real time energy management and bus voltagedroop control in solar powered standalone DC microgrid, *TENSYMP 2017 - IEEE International Symposium on Technologies for Smart Cities*, 2017, doi: 10.1109/TENCONSpring.2017.8070056.

[23] R. Sabzehgar, A review of AC/DC microgrid-developments, technologies, and challenges, in *2015 IEEE Green Energy and Systems Conference, IGESC 2015*, 2015, pp. 11–17, doi: 10.1109/IGESC.2015.7359384.

[24] F. Nejabatkhah and Y. W. Li, Overview of power management strategies of hybrid AC/DC microgrid, *IEEE Transactions on Power Electronics*, 30, 12, pp. 7072–7089, 2015, doi: 10.1109/TPEL.2014.2384999.

[25] D. Zammit, et al., Overview of buck and boost converters modelling and control for stand-alone DC microgrid operation, *Offshore Energy & Storage Symposium (OSES 2016), Valletta, Malta*. Vol. 2294, 2016.

[26] W. Jing, C. H. Lai, S. H. W. Wong, and M. L. D. Wong, Battery-supercapacitor hybrid energy storage system in standalone DC microgrids: A review, *IET Renew. Power Gen.*, 11, 4. pp. 461–469, 2017, doi: 10.1049/iet-rpg.2016.0500.

[27] A. Parida and D. Chatterjee, Stand-alone AC-DC microgrid-based wind-solar hybrid generation scheme with autonomous energy exchange topologies suitable for remote rural area power supply, *Int. Trans. Electr. Energy Syst.*, 28, 4, 2018, doi: 10.1002/etep.2520.

[28] M. Sechilariu, B. C. Wang, and F. Locment, Supervision control for optimal energy cost management in DC microgrid: Design and simulation, *Int. J. Electr. Power Energy Syst.*, 58, pp. 140–149, 2014, doi: 10.1016/j.ijepes.2014.01.018.

[29] T. Morstyn, B. Hredzak, G. D. Demetriades, and V. G. Agelidis, Unified distributed control for DC microgrid operating modes, *IEEE Trans. Power Syst.*, 31, 1, pp. 802–812, 2016, doi: 10.1109/TPWRS.2015.2406871.

[30] I. Tank and S. Mali, Renewable based DC microgrid with energy management system. *2015 IEEE International Conference on Signal Processing, Informatics, Communication and Energy Systems, SPICES 2015*. 2015. doi: 10.1109/SPICES.2015.7091542.

7 Blockchain and Smart Grid

Satpal Singh Kushwaha, Sandeep Joshi, Amit Kumar Bairwa, and Sandeep Chaurasia
Manipal University

CONTENTS

7.1 INTRODUCTION

Blockchain is gaining its popularity day by day and growing at a very fast pace. It can be defined as a decentralized distributed ledger technology, which records transactions in an immutable way. It means a transaction that has been recorded on the blockchain will not be altered anyhow. Due to this characteristic of blockchain (Zheng et al. 2017), it is gaining popularity day by day, because it eliminates the need of third parties in the transaction processing in between two persons. The distributed and decentralized nature of blockchain technology increases the transparency in the transactions. So now two unknown persons at the very distant parts of the worlds can make deals of digital assets without requiring a trusted entity.

7.2 BLOCKCHAIN

Blockchain (Hewa, Ylianttila, and Liyanage 2021) is a collection of interconnected nodes with the help of cryptographic hash; each block of the chain contains several transactions. So, every time when a new transaction is recorded, the same is added to all the participants' ledger, because this distributed database is managed by several different participants across the network. So, a blockchain has several features such as programmable, distributed, decentralized, immutable, timestamped, unanimous, anonymous, and secure. These features make a blockchain immutable. A block in

DOI: 10.1201/9781003272717-7

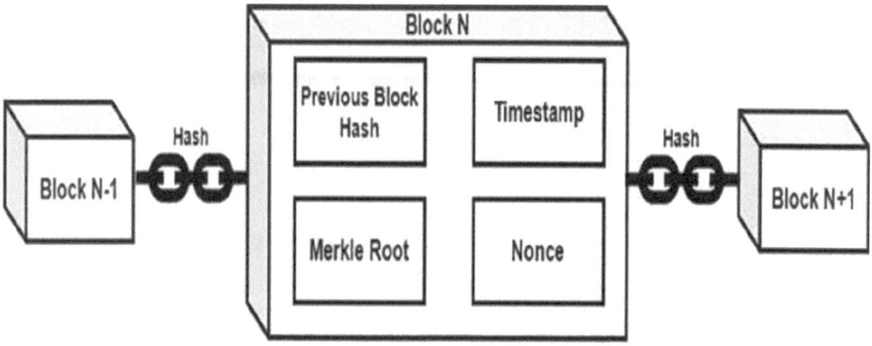

FIGURE 7.1 Structure of blockchain nodes.

the blockchain contains several important pieces of information such as Merkle root, nonce, timestamp, world state, and previous block hash. Figure 7.1 depicts the blocks structure of a blockchain.

The blockchain is a decentralized distributed ledger. It is decentralized because it is not managed by a single entity, but all the nodes of the blockchain. It is distributed because the data are not stored at a single place, but they are stored at all the nodes of the blockchain. So, this makes the blockchain transparent and very much attracting.

7.2.1 Blockchain Categories

a. Permissioned vs Permissionless

The main difference in between the permissioned and permissionless blockchain is related to joining the node creation in the blockchain. In permissioned blockchains, only the authorized person can do the same, but in the permissionless blockchain, anyone can do it.

b. Private vs Public

In case of public blockchain, everything is open and decentralized, where any one can view or validate the nodes. In contrast, in the case of private blockchain, complete decentralization is not there; in other words, the same is partially decentralized in nature.

c. On-chain vs Off-chain

It can be defined as the transactions in the blockchain are available on the blockchain network either on-chain or off-chain in terms of visibility. The off-chain are more popular due to the low cost and less execution time.

7.2.2 Features of Blockchain (Zheng et al. 2017)

a. Decentralization

It means there is no single centralized authority to control all the operations in the blockchain, but it is supported by all nodes in network. It is done by using consensus protocols such as proof of Byzantine, proof of work, and proof of stake.

b. Scalability

Blockchain is scalable in nature because every time a new transaction is executed, a new node is included to the blockchain. In blockchain, scalability is achieved by the distributed nature of the blockchain.

c. Trustless, but Secure

Here, no trusted third party is required to validate the operations, but it is performed in a peer-to- peer manner. The trust is maintained by the consensus protocols. So, the blockchain removes the requirement of the trusted third party.

d. Immutability

This is one of the extremely crucial features of the blockchain that make the data on the chain secure by making them unalterable. It means once any transaction is deployed on the blockchain, then no change can be possible after that.

e. Transparency and Auditability

The decentralized feature of the blockchain makes all the data available to all the nodes in the network, which can be verifiable by anyone. This makes it transparent and auditable.

f. Secure Digital Contracts

These are secure auto-executable scripts known as smart contracts, which are auto-executable contract codes and get executed when certain conditions are triggered. These smart contracts are deployed on the blockchain and contain the agreement terms in the coded scripts. These scripts do not require any human intervention, but are executed based on predefined terms and conditions. And as the smart contracts are deployed on the blockchain, these are immutable in nature, which make them perfect.

7.3 SMART GRID

Due to societal and technical developments, it is not possible to satisfy the energy demands from the traditional electricity generation sources. World's population is increasing gradually; in the same way, the need of electricity as an energy source is rising in a very fast speed. The following graph shows the global electricity consumption since 1980–2018 (Figure 7.2).

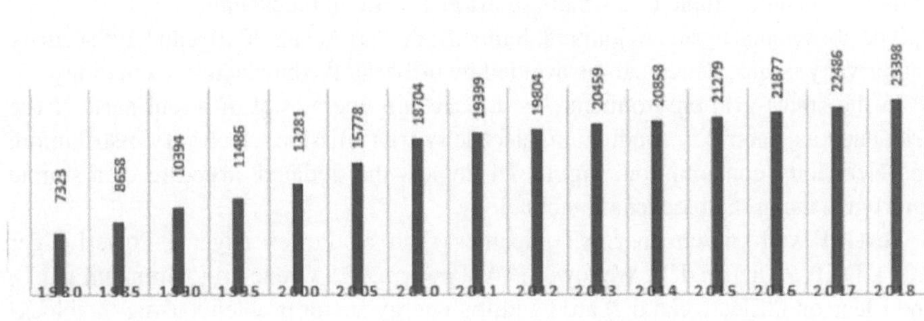

FIGURE 7.2 Global electricity consumption since 1980–2018 (Alves 2021).

So, to fulfill this increased energy demand, the renewable energy sources can play a very crucial role. The solar energy and wind energy are the key players with highest stake in the field of renewable energy. Traditional and renewable energy sources are very much different from each other in terms of characteristics for distributing the energy. In the traditional energy sources, there is a transmission loss during transit. The smart grid (Munsing, Mather, and Moura 2017) can eliminate these losses by integrating these distributed energy sources. But the centralized management of the smart grid is very costly and poses security threats to the electricity grid. So, the decentralized nature of the blockchain can resolve this issue.

7.4 BLOCKCHAIN IN SMART GRID

In traditional power grids, all the operations are managed by the central authority. But traditional power grids fail to fulfill the consumer requirements. The smart grid integrates the distributed energy sources for filling the gap of electricity supply to the consumers. So, the centralized power grids authorities can be removed by simply converting the same to decentralized smart grid (Aderibole et al. 2020). But this decentralization poses security (Alam, Islam, and Ferdous 2019; Sun, Hahn, and Liu 2018) and privacy threats to the transactions (Yu et al. 2019) in the smart grid, such as unauthorized access or modification in the transaction data. These issues can be resolved by using the revolutionary blockchain technology in smart grid (Musleh, Yao, and Muyeen 2019). All the features of the blockchain make the smart grid an excellent power grid with no issue in security, privacy, and trust. In a traditional smart grid, three entities exist to manage the whole system, which are as follows:

i. Producer: This is the one who generates and transmits the electricity to others.
ii. Consumer: This is the one who only consumes the electricity.
iii. Producer-Consumer: This is the one who perform both the tasks, which means generation and consumption according to requirements. This one can be anyone such as a house or a factory or a school or a government office.

Figure 7.3 is an example of a simple smart grid without blockchain.

The above smart grid (Arjomand, Sami Ullah, and Aslam 2020) can have security and privacy issues, which can be avoided by utilizing the blockchain technology.

In the smart grid environment, the blockchain operates at different parts of the grid, such as electricity production, electricity transmission, electricity distribution, and electricity consumption. Figure 7.4 depicts the updated structure of a simple smart grid using the blockchain technology.

Several well-known energy companies such as PowerLedger ("PowerLedger n.d."), Drift ("Drift n.d."), WePower ("WePower n.d."), Greneum ("Greneum n.d."), and Electron ("Electron n.d.") are handling energy sector problems using the blockchain technology.

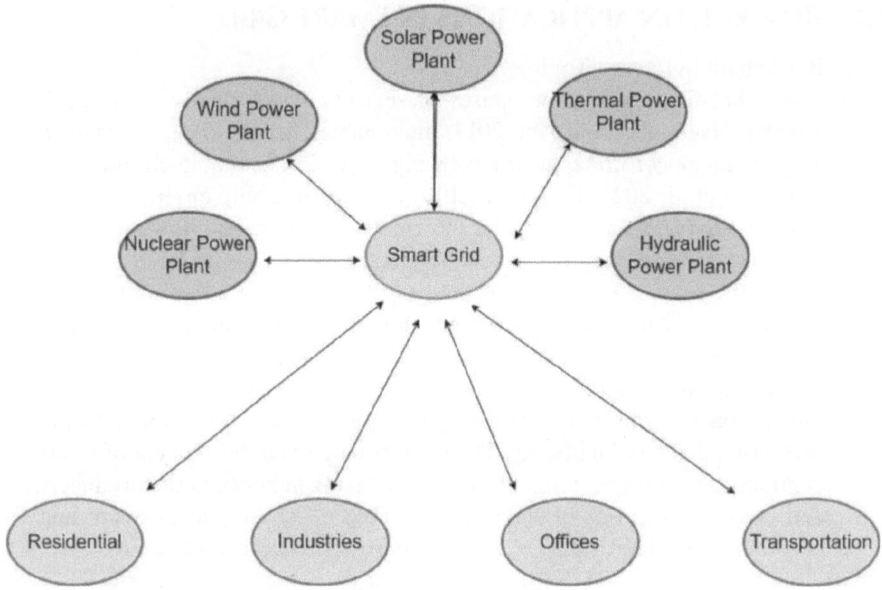

FIGURE 7.3 Example of a simple smart grid.

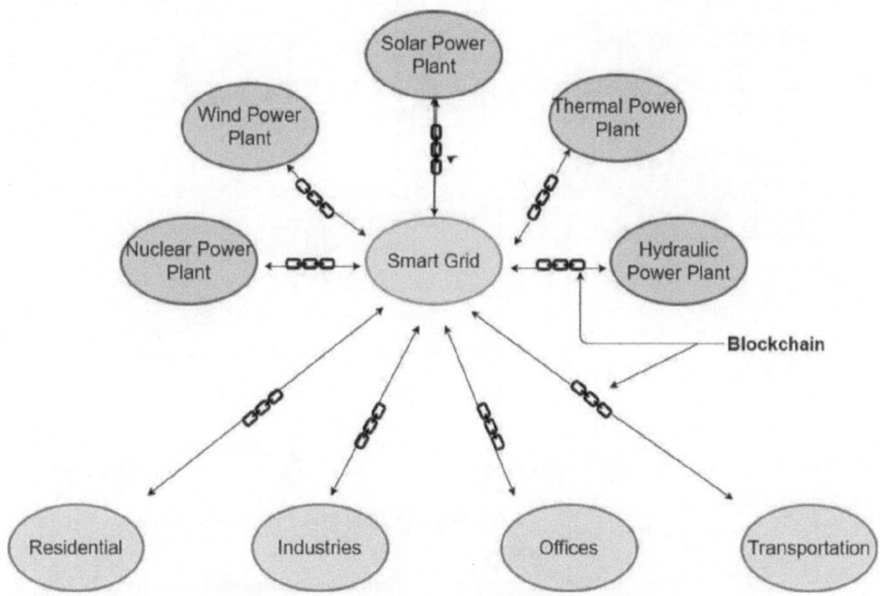

FIGURE 7.4 Example of a simple smart grid with blockchain.

7.5 BLOCKCHAIN APPLICATIONS IN SMART GRID

a. Blockchain in Energy Trading

Blockchain removes the causes of security flaws (Wang et al. 2019; Tanaka, Nagakubo, and Abe 2017) and enables the *producer*, *consumer*, and *producer-consumer* to trade (Kang et al. 2017; Sabounchi and Wei 2017; Liu et al. 2021; Luo et al. 2019) directly in a safe environment. In case of *producer-consumer*, the information and energy flows are bidirectional because they can consume electricity in case of need and can supply the surplus electricity to the grid. So, this direct trading removes the need for a trusted third party, which is required in the traditional grid operation to act as an intermediary and allows indirect communication in between *producer* and *consumer*. Li et al. (2018) proposed an energy trading approach in the smart grid using the consortium blockchain to ensure protection and confidentiality. The authors employed the concept of smart contracts for auto-execution of the agreed terms in between the producers and consumers, proof of stake for ensuring consensus mechanism, and cryptographic algorithms for security. Figure 7.5 depicts the same.

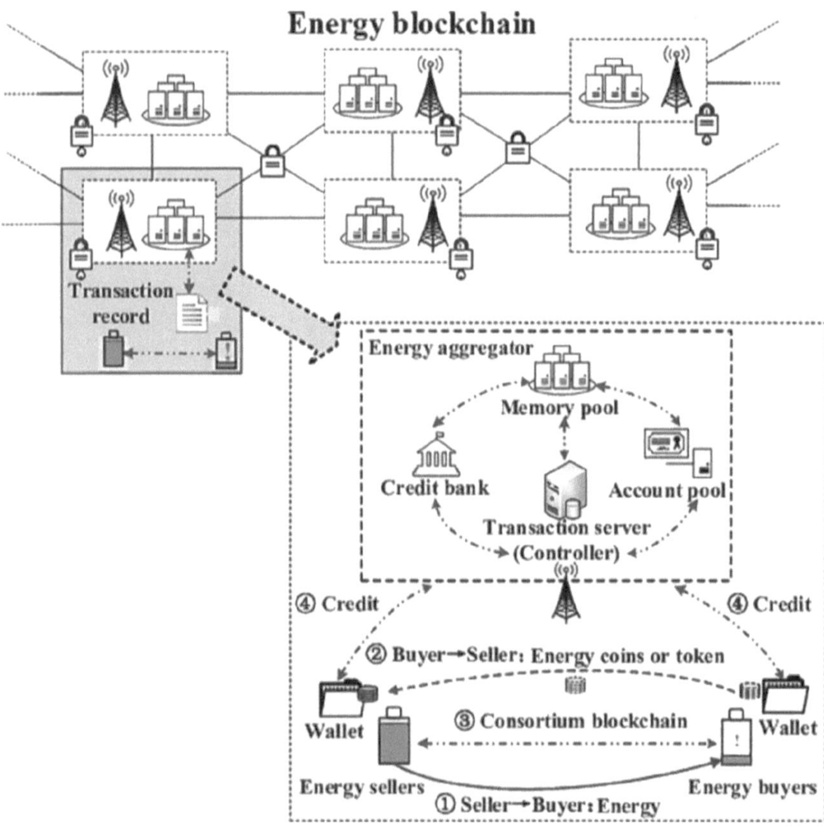

FIGURE 7.5 Consortium blockchain-based secure energy trading system (Li et al. 2018).

b. Blockchain Applications in Electric Vehicles

The electric vehicles can be considered as a part of smart grid because the charging of the electric vehicles in some uncoordinated way may increase load on the main power grid. So, this can be avoided by integrating the blockchain technology in this. Several approaches have been proposed for integrating the electric vehicles with the blockchain technology for reducing power fluctuation or locating the nearest charging stations. Garg et al. (2019) proposed a systematic approach for blockchain technology integration with electric vehicles. This approach is based on elliptic curve cryptography-assisted hierarchical authentication mechanism. Figure 7.6 depicts the system of Garg et al. (2019).

c. Blockchain Applications in Microgrid Operations

In traditional energy grids, trust and security issues still exist. These issues can be removed with the help of blockchain technology. Several researchers proposed several approaches (Neagu, Grigoras, and Ivanov 2019) to integrate the microgrid operation with the blockchain technology. Wang et al. proposed an electricity transaction structure for solving the transactional management problems that exist in the traditional

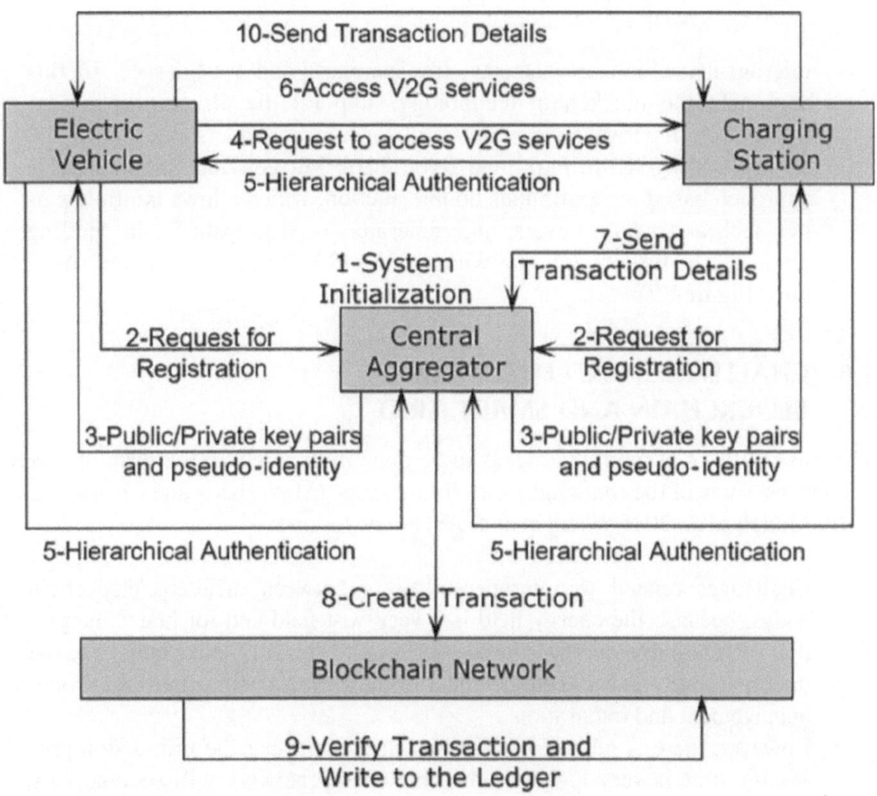

FIGURE 7.6 Integration approach for vehicle-to-grid environment (Garg et al. 2019).

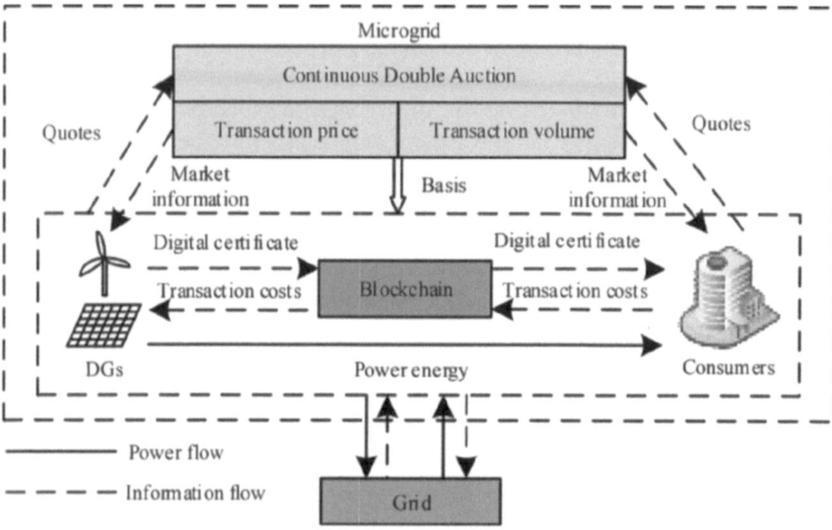

FIGURE 7.7 Electricity transaction in microgrids using blockchain (Wang et al. 2017).

microgrids, which are security, transparency, and trust issues. In this approach, the blockchain technology supports the direct peer-to-peer transaction in between generators and consumers. This is a decentralized energy trading (Kim, Park, and Ryou 2018; Muyeen and Rahman 2017) approach based on continuous double auction, which allows multiple parties such as traders, buyers, and generator to participate in the bidding process. The following is the Wang et al.'s (2017) proposed system structure (Figure 7.7).

7.6 CHALLENGES AND FUTURE OF BLOCKCHAIN AND SMART GRID

There is still a lot of work that needs to be done in this field in the form of several challenges; some of the challenges with future scope (Makridakis and Christodoulou 2019; Mollah et al. 2021; Wang et al. 2018) are as follows:

a. Challenges related to interoperability in between different blockchain nodes, because the energy field is a very vast field and for better integration of renewable energy sources will create several blockchains. So, the interoperability in between the nodes of these blockchain is must for proper management and integration.
b. However, there is no intermediary in blockchain, but the transaction processing time is very low. So, when the energy network will become very large, then this low transaction time will reduce the effectiveness of the integration, because the throughput is directly related to transaction time.

c. The selection of a good consensus mechanism. The proof of work mechanism is very costly because it requires a lot of energy for confirming a transaction. The proof of stake mechanism is having a problem of dominance of some nodes because of their stake. The proof of BFT mechanism is not suitable for very large networks where the number of participants is very high. So, the selection of a suitable mechanism from these is still a challenge that needs to be solved in future.
d. Challenges related to smart contract security. Public blockchains are more susceptible to attacks due to their openness. Smart contracts are a part of blockchain and contain scripts, so a vulnerable script or error can be the reason for a loss of millions. These issues should be handled in future.

7.7 CONCLUSIONS

Blockchain's adaptability is increasing day by day in different fields, which is due to its excellent characteristics. Several startups and some well-known companies of energy sector have already started using the blockchain technology in smart grid. But several challenges still exist related to transaction time, cost of transaction, etc. These significant challenges should be countered in near future. But we cannot deny the thing that the integration of blockchain technology in smart grid will be revolutionary in energy sector because the utilization of blockchain technology in power sector is incredibly impressive.

REFERENCES

Aderibole, Adedayo, Aamna Aljarwan, Muhammad Habib Ur Rehman, Hatem H. Zeineldin, Toufic Mezher, Khaled Salah, Ernesto Damiani, and Davor Svetinovic. 2020. Blockchain technology for smart grids: Decentralized NIST conceptual model. *IEEE Access* 8: 43177–90. https://doi.org/10.1109/ACCESS.2020.2977149.

Alam, Asraful, Mohammad Tausiful Islam, and Arafa Ferdous. 2019. Towards blockchain-based electricity trading system and cyber resilient microgrids. In *2019 International Conference on Electrical, Computer and Communication Engineering (ECCE)*, 1–5. IEEE. https://doi.org/10.1109/ECACE.2019.8679442.

Alves, Bruna. 2021. *Net Consumption of Electricity Worldwide in Select Years from 1980 to 2018*. https://www.statista.com/statistics/280704/world-power-consumption/.

Arjomand, N., H. Sami Ullah, and S. Aslam. 2020. *A Review of Blockchain-Based Smart Grid: Applications, Opportunities, and Future Directions, no. 2002.05650*. https://arxiv.org/abs/2002.05650v2.

Drift. n.d. https://www.joindrift.com.

Electron. n.d. https://electron.net.

Garg, Sahil, Kuljeet Kaur, Georges Kaddoum, Francois Gagnon, and Joel J.P.C. Rodrigues. 2019. An efficient blockchain-based hierarchical authentication mechanism for energy trading in V2G environment. *2019 IEEE International Conference on Communications Workshops, ICC Workshops 2019- Proceedings*, no. May 20–24. https://doi.org/10.1109/ICCW.2019.8756952.

Greneum. n.d. https://greeneum.net.

Hewa, Tharaka, Mika Ylianttila, and Madhusanka Liyanage. 2021. Survey on blockchain based smart contracts: Applications, opportunities and challenges. *Journal of Network and Computer Applications* 177 (March): 102857. https://doi.org/10.1016/j.jnca.2020.102857.

Kang, Jiawen, Rong Yu, Xumin Huang, Sabita Maharjan, Yan Zhang, and Ekram Hossain. 2017. Enabling localized peer-to-peer electricity trading among plug-in hybrid electric vehicles using consortium blockchains. *IEEE Transactions on Industrial Informatics* 13 (6): 3154–64. https://doi.org/10.1109/TII.2017.2709784.

Kim, GeunYoung, Junhoo Park, and Jaecheol Ryou. 2018. A study on utilization of blockchain for electricity trading in microgrid. In *2018 IEEE International Conference on Big Data and Smart Computing (BigComp)*, 743–46. IEEE. https://doi.org/10.1109/BigComp.2018.00141.

Li, Zhetao, Jiawen Kang, Rong Yu, Dongdong Ye, Qingyong Deng, and Yan Zhang. 2018. Consortium blockchain for secure energy trading in industrial internet of things. *IEEE Transactions on Industrial Informatics* 14 (8): 3690–700. https://doi.org/10.1109/TII.2017.2786307.

Liu, Chao, Kok Keong Chai, Xiaoshuai Zhang, and Yue Chen. 2021. Peer-to-peer electricity trading system: Smart contracts based proof-of-benefit consensus protocol. *Wireless Networks* 27, 4217–28. https://doi.org/10.1007/s11276-019-01949-0.

Luo, Fengji, Zhao Yang Dong, Gaoqi Liang, Junichi Murata, and Zhao Xu. 2019. A distributed electricity trading system in active distribution networks based on multi-agent coalition and blockchain. *IEEE Transactions on Power Systems* 34 (5): 4097–108. https://doi.org/10.1109/TPWRS.2018.2876612.

Makridakis, Spyros, and Klitos Christodoulou. 2019. Blockchain: Current challenges and future prospects/applications. *Future Internet* 11 (12): 258. https://doi.org/10.3390/fi11120258.

Mollah, Muhammad Baqer, Jun Zhao, Dusit Niyato, Kwok Yan Lam, Xin Zhang, Amer M.Y.M. Ghias, Leong Hai Koh, and Lei Yang. 2021. Blockchain for future smart grid: A comprehensive survey. *IEEE Internet of Things Journal* 8 (1): 18–43. https://doi.org/10.1109/JIOT.2020.2993601.

Munsing, Eric, Jonathan Mather, and Scott Moura. 2017. Blockchains for decentralized optimization of energy resources in microgrid networks. In *2017 IEEE Conference on Control Technology and Applications (CCTA)*, 2164–71. IEEE. https://doi.org/10.1109/CCTA.2017.8062773.

Musleh, Ahmed S., Gang Yao, and S. M. Muyeen. 2019. Blockchain applications in smart grid-review and frameworks. *IEEE Access* 7: 86746–57. https://doi.org/10.1109/ACCESS.2019.2920682.

Muyeen, S. M. Muyeen, and Saifur Rahman. 2017. *Communication, Control and Security for the Smart Grid*. Institution of Engineering and Technology. https://doi.org/10.1049/PBPO095E.

Neagu, Bogdan Constantin, Gheorghe Grigoras, and Ovidiu Ivanov. 2019. An efficient peer-to-peer based blokchain approach for prosumers energy trading in microgrids. In *2019 8th International Conference on Modern Power Systems (MPS)*, 1–4. IEEE. https://doi.org/10.1109/MPS.2019.8759743.

PowerLedger. n.d. https://www.powerledger.io.

Sabounchi, Moein, and Jin Wei. 2017. Towards resilient networked microgrids: Blockchain-enabled peer-to-peer electricity trading mechanism. In *2017 IEEE Conference on Energy Internet and Energy System Integration (EI2)*, 1–5. IEEE. https://doi.org/10.1109/EI2.2017.8245449.

Sun, Chih-Che, Adam Hahn, and Chen-Ching Liu. 2018. Cyber security of a power grid: State-of-the-art. *International Journal of Electrical Power & Energy Systems* 99 (July): 45–56. https://doi.org/10.1016/j.ijepes.2017.12.020.

Tanaka, Kenji, Kosuke Nagakubo, and Rikiya Abe. 2017. Blockchain-based electricity trading with digitalgrid router. In *2017 IEEE International Conference on Consumer Electronics – Taiwan (ICCE-TW)*, 201–2. IEEE. https://doi.org/10.1109/ICCE-China.2017.7991065.

Wang, Boyu, Morteza Dabbaghjamanesh, Abdollah Kavousi-Fard, and Shahab Mehraeen. 2019. Cybersecurity enhancement of power trading within the networked microgrids based on blockchain and directed acyclic graph approach. *IEEE Transactions on Industry Applications* 55 (6): 7300–9. https://doi.org/10.1109/TIA.2019.2919820.

Wang, Jian, Qianggang Wang, Niancheng Zhou, and Yuan Chi. 2017. A novel electricity transaction mode of microgrids based on blockchain and continuous double auction. *Energies* 10 (12): 1–22. https://doi.org/10.3390/en10121971.

Wang, Shuai, Yong Yuan, Xiao Wang, Juanjuan Li, Rui Qin, and Fei-Yue Wang. 2018. An overview of smart contract: Architecture, applications, and future trends. In *2018 IEEE Intelligent Vehicles Symposium (IV)*, 108–13. IEEE. https://doi.org/10.1109/IVS.2018.8500488.

WePower. n.d. https://wepower.com.

Yu, Yunjun, Yanghui Guo, Weidong Min, and Fanpeng Zeng. 2019. Trusted transactions in micro-grid based on blockchain. *Energies* 12 (10): 1952. https://doi.org/10.3390/en12101952.

Zheng, Zibin, Shaoan Xie, Hongning Dai, Xiangping Chen, and Huaimin Wang. 2017. An overview of blockchain technology: Architecture, consensus, and future trends. In *2017 IEEE International Congress on Big Data (BigData Congress)*, 557–64. IEEE. https://doi.org/10.1109/BigDataCongress.2017.85.

8 Renewable Energy Source Technology with Geo-Spatial-Based Intelligent Vision Sensing and Monitoring System for Solar Aerators in Fish Ponds

K. Sujatha
Dr. M.G.R. Educational and Research Institute

N.P.G. Bhavani
Saveetha School of Engineering

K. Krishnakumar
Vels institute of Science, Technology & Advanced Studies

U. Jayalatsumi, T. Kavitha, and K. Senthil Kumar
Dr. M.G.R. Educational and Research Institute

CONTENTS

DOI: 10.1201/9781003272717-8

8.1 INTRODUCTION

A huge amount of emissions are liberated from thermal power plants, which contain carbon dioxide (CO_2), sulphur dioxide (SO_2), and carbon monoxide (CO) that pollute the atmosphere, and there is a depletion in the natural resources such as fossil fuels. The researchers' main motivation has been to contribute to renewable energy technology. Solar energy, which is a renewable form of energy that is abundant in India for around 300days a year, has piqued the curiosity of many academics in this area. Hence, the researchers have decided to use the solar energy to power the aerators associated with the solar pond. The solar aerators play an important role in aquaculture. The basic concept addressed here is to maximize solar energy output by constructing a cost-effective and specific solar tracking system. The real problem is to achieve optimal energy output yield. Certain solar systems using concentrated optical systems such as concentrated photovoltaic (CPV) systems have consistently outperformed the usual type of photovoltaic (PV) systems in terms of energy efficiency. The solar systems using optical system consists of fixed and tuneable mirror strips and also convex lens systems to focus the beam of light at 90° with respect to the surface area of the panel to expand the output of solar energy. As a result, a high-precision, effective tracking system is required. Existing tracking algorithms that use optical or brightness sensors are inaccurate since they can't tell whether the sun is shining or not. Figure 8.1 depicts the schematic for a solar-powered aerator.

8.2 LITERATURE REVIEW

8.2.1 International Status

The solar tracking system, which consists of an electrical device attached to an electro-mechanical assembly that tracks the sun's position in relation to a reference [1–9], was addressed by the researchers. Two dual-axis DC motors with a slew drive is used to track the angle of azimuth and zenith. A webcam is utilized in tracking the sun by sensing the light radiation to analyse the tracking error. A pipe tool is used to investigate the display locally. A low-cost embedded system using microprocessor is used for controlling and monitoring the solar system. Mexico is a country where this kind of work is carried out. In their research, the Koreans revealed [10–14] how they employed a two-axis approach to follow the sun. A GPS and a camera sensor are used

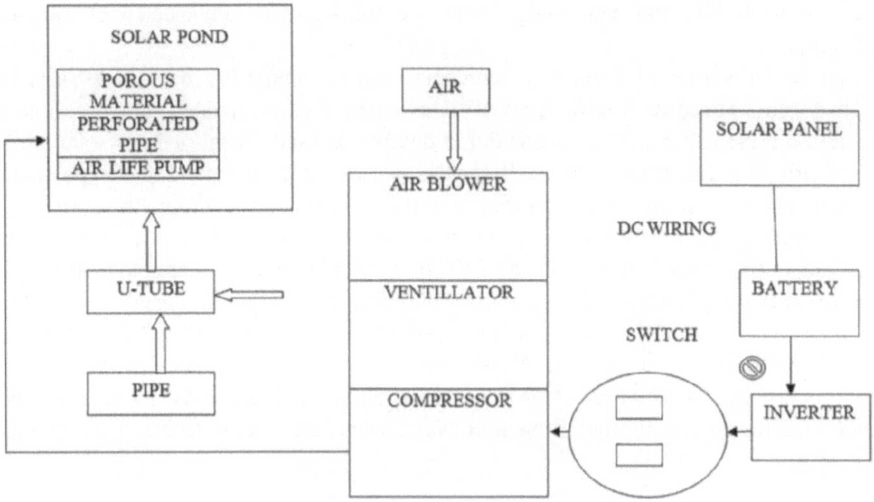

FIGURE 8.1 Block diagram – solar aerators.

to track the horizontal and vertical angles of the sun. Time, latitude, and longitude positions are collected from the GPS. An astronomical formula can be used to calculate the sun's current azimuth and altitude angles using this information. The present heading angle of the solar tracking system, on the other hand, may not be accurate, and it cannot be updated using GPS data. As a result, the camera sensor is employed in our research to more precisely estimate the position of the sun. The sun's centre position is calculated using image processing, and the solar panel is shifted to find the sun's centre point on image centre.

Throughout the year, the Malaysian country is drenched in sunlight. This is an added benefit for them because they can use solar energy by employing a webcam and image processing algorithm incorporated in an Arduino board. The authors in Ref. [15,16] proposed this approach. Jordanian researchers concluded [17,18] that the prediction of the sun's location throughout the summer, cloudy, dusty, and rainy seasons is done using filtering algorithms and segmented for correct and consistency in focal point detection using self-identification techniques. The data pattern is studied by the ANN, wherein the difference between the actual and target data is used for error calculation and modification. The scientist Taoyuan [19,20] also deployed a sensor that not only captures the image of the sun, but also indicates the location of the sun using a position sensor. The colour images of the sun are converted to greyscale values so as to detect the presence of noise that affects the quality.

8.2.2 National Status

Investigators have designed solar cells that are of low cost with improved efficacy. The photovoltaic array in a solar panel consists of solar cells, which converts the light energy of the sun into electrical energy. The efficiency of such solar cells is found to be 15%. Innovations are taking place continuously in this domain to improve the extraction of light and heat energy from the sun by using advanced technologies [21–24].

At the University of Toronto, researchers used nanoparticles that are sensitive to light for this purpose. A novel kind of light-sensitive nanoparticles named colloidal quantum dots, which are economical and flexible, is used for fabricating solar cells. Explicitly, the use of new materials for the fabrication of N-type and P-type semiconductors can improve the performance of the solar cells. This unique invention has alleviated the previously existing problems, and therefore, the design modernization has led to the transformation in the field of renewable energy technology to use it for real-time applications. The scientists observed that n-type materials connect to oxygen, whereas these novel colloidal quantum dots do not bind to air and hence may keep their stability outside. Hence, this characteristic enhances the absorption of light energy. The solar panels that are incorporated with this novel technique yield increased output during the efficient conversion of light energy to electrical energy [25–30].

8.3 NOVELTY OF PROPOSED WORK

The novelty of this technology is that a vision camera is used for solar image acquisition from the sun so as to track its position. The cost of camera sensors and their associated data processing devices has decreased in recent years, which has attracted the interest of academics in this field. The main theme behind this work is that the light intensity of the sun is directly proportional to the electrical energy that is produced, which is in turn dependent on the weather conditions. The sun tracking algorithm equipped in the central processing unit depends on the weather data stored in cloud storage, which is displayed simultaneously. This value-added information enables to enhance the accuracy of the solar system, thereby yielding an optimal output.

8.4 OBJECTIVES

The prime objectives behind this work are as follows:

a. The control scheme proposed here is a feed forward sun tracking algorithm using image processing techniques.
b. The efficiency of the solar panel design is inferred by implementing of IPT, WOPO, and SSPO algorithms at maximum peak and sun tracking controller for precise sun tracking.
c. Geo-spatial information system for monitoring and control of the output parameters without interruption.
d. To offer an inexpensive and sophisticated result by consuming less power to the solar aerators associated with fish farming.

The tree diagram in Figure 8.2 describes the scope of this scheme. Various algorithms and techniques for extracting the features from the tracked images of the sun are used to yield maximum power output from the solar panel. This technology is based on solar energy. The site of implementation of the solar panel requires

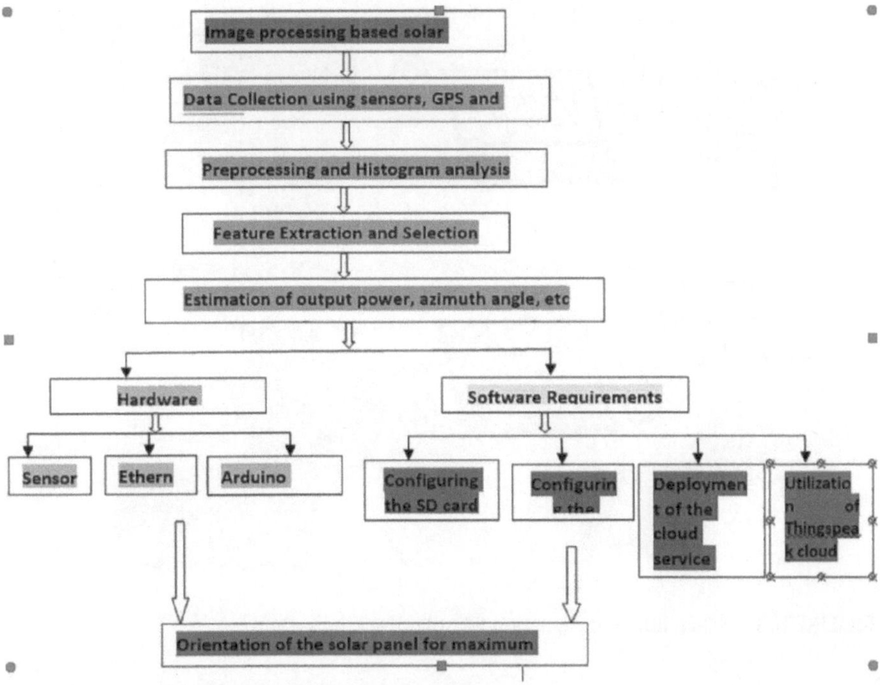

FIGURE 8.2 Schematic for the scope of the work.

open space exposed to sun's radiation. The solar energy incident on the solar panel exposed to atmosphere is converted into electrical energy, which is used by the solar aerators to supply oxygen to the solar pond for sustaining the aquaculture.

8.5 METHODOLOGY

Figure 8.3 depicts the schematic of the proposed solar tracking system with the goal of maximizing production. Solar energy, for example, plays a critical role in the power business, helping to meet the expanding demands for electricity in the industrial sector and other utilities (Table 8.1). In this chapter, solar energy is utilized for operating solar aerators to supply oxygen for enhancing fish farming in solar ponds. The power businesses, particularly those based on renewable energy sources, have grown dramatically in recent years. The rapid growth of the wind power business has also resulted in several issues being noted. The work methodology includes the data collection and images of the sun with its geo-spatial information. For simulation study, MATLAB is used. The block diagram for MPPT is denoted in Figure 8.4.

- Use GPS to get the satellite data.
- Create the corrective angle initially.
- Using an image sensor, obtain the sun's coordinates.

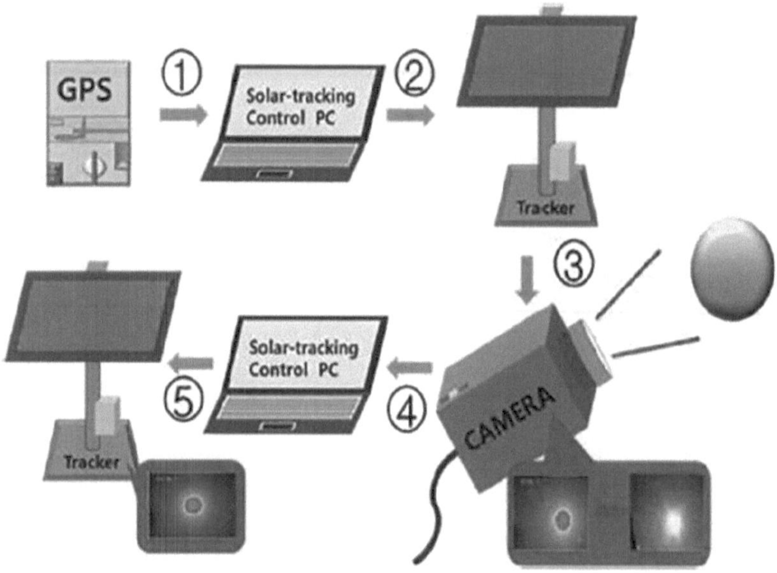

FIGURE 8.3 Schematic representation for solar tracking system.

- Use an image sensor to get weather data.
- Once again, create the corrective angle for the second time.

Figure 8.5 shows the standardized properties of a solar panel's V–I and P–V characteristics.

8.6 PREPROCESSING

8.6.1 Edge Detection

As a part of preprocessing, the extracted images are converted into greyscale. Then noise removal is done using the Sobel operator.

% MATLAB code for preprocessing

```
im = imread('1.png');
image (im);
gray = (0.2989 * double(im(:, :, 1)) + 0.5870 * double(im(:,
:, 2)) + 0.1140 * double(im(:, :, 3)))/255;
edgeIm = sobel_mex(gray, 0.7);
imshow (edgeIm);
```

8.6.2 Histogram Analysis

The histogram is defined as the number of pixels with respect to the occurrence.

% MATLAB code for histogram analysis

TABLE 8.1
Data Collection Corresponding to Sun's Position for Various Weather Conditions

S. No.	Weather Conditions	Images of Sun	Latitude	Longitude	Azimuth Angle	Altitude	Output Power (W/m²)
1.	Sunny weather		–34	140	145.2	–55.24	600
2.			–34	140	112.31	–35.64	600
3.			–34	140	92.88	–11.43	600
4.		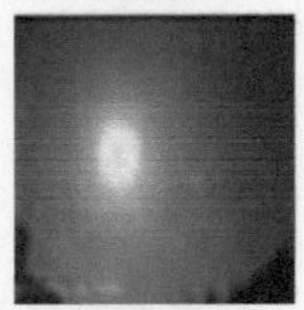	–34	140	75.79	13.23	600
5.		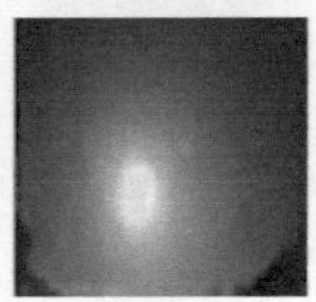	–34	140	53.87	35.74	600

(Continued)

TABLE 8.1 (*Continued*)
Data Collection Corresponding to Sun's Position for Various Weather Conditions

S. No.	Weather Conditions	Images of Sun	Latitude	Longitude	Azimuth Angle	Altitude	Output Power (W/m²)
6.			–34	140	17.58	50.57	600
7.	Cloudy weather		–34	140	146.27	–45.76	400
8.			–34	140	116.99	–26.99	400
9.			–34	140	97.82	–3.3	400
10.			–34	140	80.73	21.48	400

(*Continued*)

TABLE 8.1 (*Continued*)
Data Collection Corresponding to Sun's Position for Various Weather Conditions

S. No.	Weather Conditions	Images of Sun	Latitude	Longitude	Azimuth Angle	Altitude	Output Power (W/m²)
11.			–34	140	57.88	44.73	400
			–34	140	14.27	59.87	400
12.	Rainy weather		–34	140	17.6	70.91	200
13.			–34	140	18.15	71.48	200
14.			–34	140	18.76	72.02	200

(*Continued*)

TABLE 8.1 (*Continued*)
Data Collection Corresponding to Sun's Position for Various Weather Conditions

S. No.	Weather Conditions	Images of Sun	Latitude	Longitude	Azimuth Angle	Altitude	Output Power (W/m²)
15.			–34	140	19.42	72.53	200
16.			–34	140	20.14	73.02	200
17.			–34	140	20.92	73.49	200

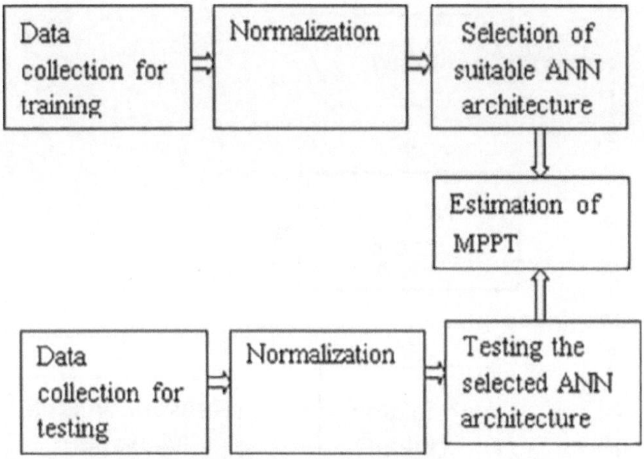

FIGURE 8.4 Block diagram for estimating the MPPT.

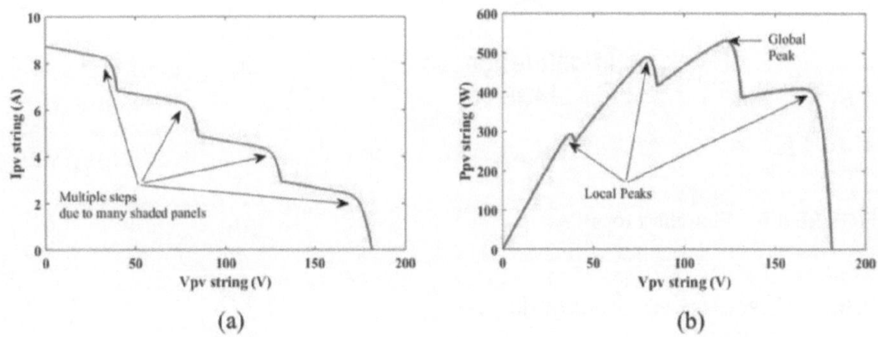

FIGURE 8.5 Solar panel V–I and P–V parameters are standard representation.

```
I = imread('1.tif');
imhist(I)
I = gpuArray(imread('1.tif'));
imshow (I);
```

8.6.3 Extraction of Features

The features are the basic patterns available in the images that get repeated for many times in different directions. The various features extracted include area, average intensity, centroid about X, centroid about Y, and orientation about X and Y using ImageJ.

8.6.4 Classification

The classification is carried out using a feed forward neural network (FFNN) trained with backpropagation algorithm (BPA). The flowchart is shown in Figure 8.6.

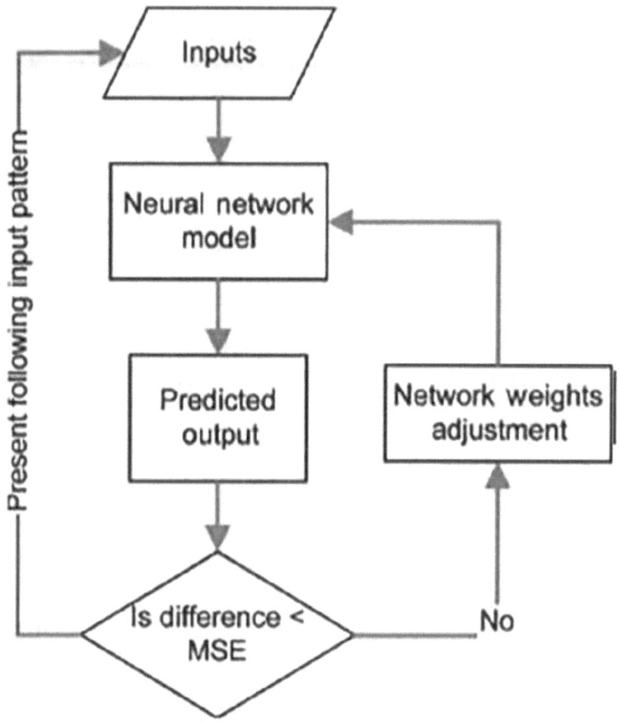

FIGURE 8.6 Flowchart for BPA.

8.6.5 Circuitry for Hardware

The Arduino is used for the implementation of hardware. A Wi-Fi connection will be enabled for the related data transfer. The reliable working condition is used to monitor the motor, which is operated to adjust the direction of the solar panel; hence, the incident radiation is always perpendicular to surface of the panel.

This scheme uses a dual-wire system with 5V supply. It is also provided with ground connection. When the incident radiation from the sun is not perpendicular to the surface of the panel, there is no conduction between the wires and the voltage at the microcontroller will go high indicating the incident light energy condition on the solar panel so that the motor will switched ON. On the other hand, when the incident radiation is perpendicular to the surface of the panel, conduction will take place due to the incident energy being converted to free electrons, which causes current to flow, and the voltage at the microcontroller will go low, indicating that the sun's incident radiation is orthogonal to the panel's surface so that the motor will be in OFF position.

8.7 RESULTS AND DISCUSSION

Algorithms using image processing techniques perform few operations on an image, which is expressed as a matrix containing the greyscale values which ranges from 0 to 255. For a colour image, these set of greyscale values get repeated in three different planes, namely the red (R), green (G), and blue (B) planes. These greyscale values are reoriented in a different fashion so as to obtain useful information by extracting the features. In this scheme, a kind of signal processing algorithm is used, which uses an image as the input, and the output is the equivalent feature values that are unique for each image. Recently, this domain has been capturing attention amidst the rapidly growing technologies. This field is the thrust area for investigative studies in the area of engineering and computer science disciplines, too.

There are three basic steps involved. They include the following:

- Use of tools for image acquisition and also to import them;
- Analysis and manipulation of the images;
- Desired output that can be flexibly altered with respect to the need.

Carbon dioxide (CO_2), sulphur dioxide (SO_2), and carbon monoxide (CO) emissions from fossil fuel power plants damage the atmosphere. Excessive consumption of fossil fuels depletes natural resources. The CO_2 and CO emissions tend to deplete the ozone layer, causing a hole in ozone layer, which in turn allows the harmful ultraviolet rays to reach the earth surface. This leads to global warming, which poses a serious problem today. The problem of global warming increases temperature on the earth surface. These ultraviolet rays causes skin allergy and respiratory problems in humans.

8.7.1 Histogram Analysis

The two types of image processing procedures used are analogue and digital image processing. Analogue image processing includes hard copies. Image analysts use a range of tools for analysing the images. Digital image processing techniques enable computer-assisted image manipulation. When adopting digital approaches, all types of data must go through three general processes: noise removal, intensification, and arrangement, including information extraction.

The histogram shows the frequency of occurrence of different pixel intensity values. On sunny days, the rate of availability of maximum intensity (255) is high in comparison with other weather conditions such as overcast and rainy days, as illustrated in Figure 8.7a. Similarly, during overcast and rainy days, Figure 8.7b and c indicates that the maximum intensity is 160 and practically equivalent to ‘0’.

8.7.2 Extraction of Features

ImageJ is used as the simulation package for extracting the features that are tabulated in Table 8.2. It is evident that features such as area, SD, and variance show a significant change for all the three categories of the captured solar images.

8.7.3 ANN-Based Classification

The parameters for ANN are recorded in Table 8.3. It’s assumed that the output power is calculated by following the sun’s images under various weather conditions. The

TABLE 8.2
Extraction of Features

S. No.	Area	Mean	SD	Mode	Min	Max	*X*	*Y*	*XM*	*YM*	Width	Height	Median
1	5550	158.039	57.013	255	5	255	37	37.5	32.666	35.527	74	75	162
2	4900	106.835	52.013	255	1	255	35	35	31.418	34.019	70	70	105
3	5475	111.49	61.959	255	0	255	36.5	37.5	32.715	37.329	73	75	102
4	5621	106.879	66.307	255	0	255	38.5	36.5	35.34	38.773	77	73	93
5	5621	101.505	68.833	255	0	255	36.5	38.5	34.199	41.549	73	77	84
6	5550	158.039	57.013	255	5	255	37	37.5	32.666	35.527	74	75	162
7	5700	127.267	60.485	255	2	255	37.5	38	32.676	35.949	75	76	126
8	4900	106.835	52.013	255	1	255	35	35	31.418	34.019	70	70	105
9	5475	111.49	61.959	255	0	255	36.5	37.5	32.715	37.329	73	75	102
10	5621	106.879	66.307	255	0	255	38.5	36.5	35.34	38.773	77	73	93
11	4830	170.425	41.409	255	26	255	35	34.5	36.014	32.165	70	69	175
12	5700	127.267	60.485	255	2	255	37.5	38	32.676	35.949	75	76	126
13	5037	115.089	55.278	255	6	255	36.5	34.5	34.049	32.274	73	69	105
14	3886	145.579	20.114	163	50	182	33.5	29	33.175	27.846	67	58	150
15	4026	120.463	19.908	140	41	152	33	30.5	33.741	28.915	66	61	124
16	4278	110.345	20.962	120	19	160	34.5	31	37.351	29.404	69	62	115
17	4216	124.52	26.966	110	18	255	34	31	35.824	28.999	68	62	124
18	4216	124.52	26.966	110	18	255	34	31	35.824	28.999	68	62	124
19	4830	170.425	41.409	255	26	255	35	34.5	36.014	32.165	70	69	175
20	5037	115.089	55.278	255	6	255	36.5	34.5	34.049	32.274	73	69	105
21	3886	145.579	20.114	163	50	182	33.5	29	33.175	27.846	67	58	150
22	4026	120.463	19.908	140	41	152	33	30.5	33.741	28.915	66	61	124
23	4900	106.835	52.013	255	1	255	35	35	31.418	34.019	70	70	105
24	4278	110.345	20.962	120	19	160	34.5	31	37.351	29.404	69	62	115

(Continued)

TABLE 8.2 (*Continied*)
Extraction of Features

S. No.	Area	Mean	SD	Mode	Min	Max	*X*	*Y*	*XM*	*YM*	Width	Height	Median
25	3886	145.579	20.114	163	50	182	33.5	29	33.175	27.846	67	58	150
26	3886	145.579	20.114	163	50	182	33.5	29	33.175	27.846	67	58	150
27	4026	120.463	19.908	140	41	152	33	30.5	33.741	28.915	66	61	124
28	4278	110.345	20.962	120	19	160	34.5	31	37.351	29.404	69	62	115
29	4216	124.52	26.966	110	18	255	34	31	35.824	28.999	68	62	124
30	4216	124.52	26.966	110	18	255	34	31	35.824	28.999	68	62	124
31	4830	170.425	41.409	255	26	255	35	34.5	36.014	32.165	70	69	175
32	5037	115.089	55.278	255	6	255	36.5	34.5	34.049	32.274	73	69	105
33	5475	111.49	61.959	255	0	255	36.5	37.5	32.715	37.329	73	75	102
34	5621	106.879	66.307	255	0	255	38.5	36.5	35.34	38.773	77	73	93
35	5621	101.505	68.833	255	0	255	36.5	38.5	34.199	41.549	73	77	84
36	5621	101.505	68.833	255	0	255	36.5	38.5	34.199	41.549	73	77	84
37	5550	158.039	57.013	255	5	255	37	37.5	32.666	35.527	74	75	162
38	5700	127.267	60.485	255	2	255	37.5	38	32.676	35.949	75	76	126

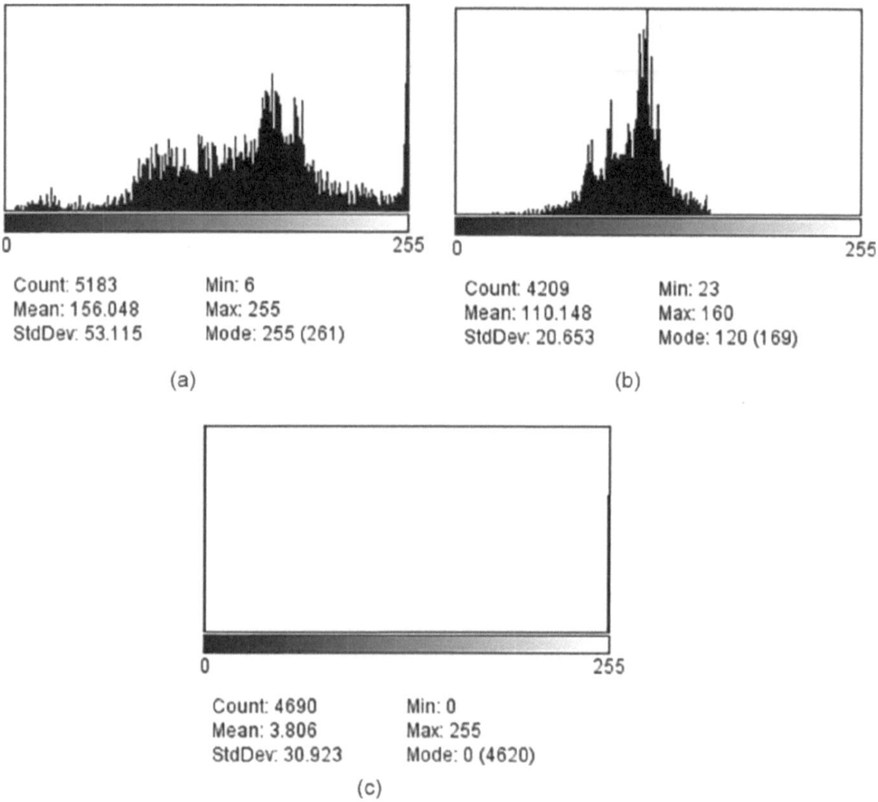

FIGURE 8.7 (a) Sunny day's histogram. (b) Cloudy day's histogram. (c) Rainy day's histogram.

TABLE 8.3
Parameters of ANN

S. No.	Performance Measure
1.	Seven nodes – input layer
2.	Four nodes – hidden layer
3.	One node – output layer
4.	Sigmoidal function – hidden layer
5.	Sigmoidal function – output layer
6.	Mean squared error is 0. 0198
7.	200 Iterations
8.	0.9 Learning factor

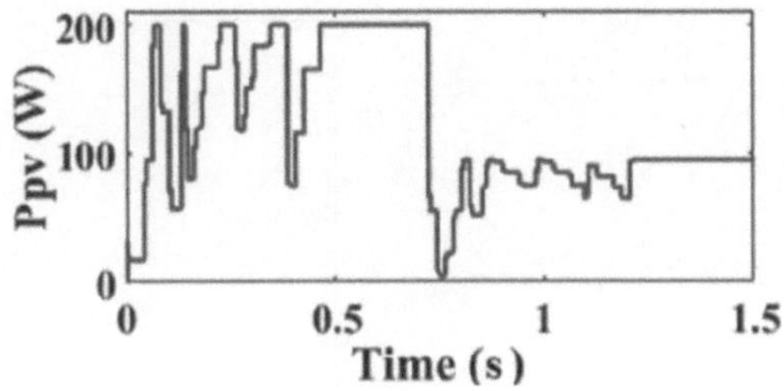

FIGURE 8.8 Tracking results for WOPO algorithm.

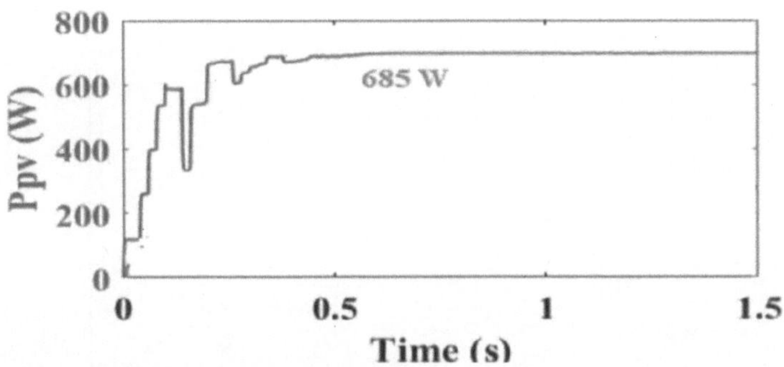

FIGURE 8.9 Tracking results for SSPO algorithm.

classification efficiency of the ANN classifier is nearer to 100%. The P–V characteristics consist of three local peaks (43.52 W, 60.63 W, and 62.95 W) and a global peak (78.26 W). The maximum power can be harvested only when the MPPT algorithm tracks the global peak (78.26 W) as in Figure 8.8. As expected, the hybrid WOPO algorithm has tracked the exact global peak power 78.2 W within four iterations for better comparison.

The single-sensor SSPO algorithm tracks the maximum power by sensing the output current, hence for better understanding the power current (P–I) characteristics. The proposed single-sensor algorithm has tracked the exact maximum input power (685 W) with less number of oscillations. The maximum output current supplied to the battery at steady state is 7.04 A (Figure 8.9).

8.7.4 Summary

The main idea is to create an intelligent classifier that can monitor and control the solar panel's power output, which has successfully been replicated here. The BPA classifier outperforms other standard classifiers in terms of robustness. According to the performance requirements, the suggested classifier was able to recall all of the photographs of the sun at various inclinations with precision. The categorization efficiency is around 100%. Figure 8.10a–c illustrates the output for classification.

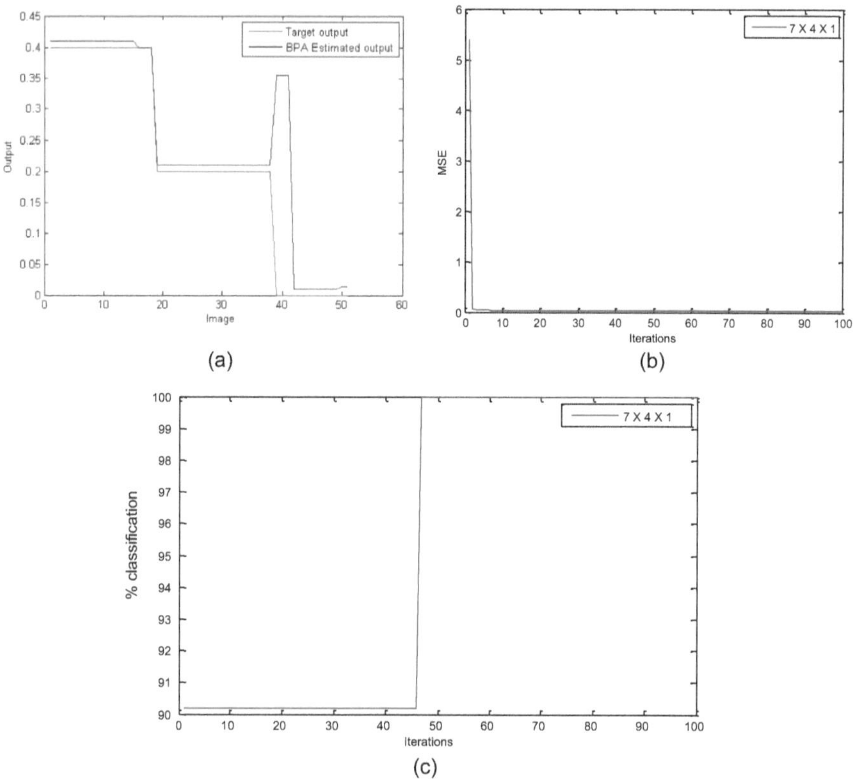

FIGURE 8.10 (a) The solar tracking system power output. (b) MSE vs. number of iterations. (c) Efficiency of classification.

8.8 CONCLUSIONS AND FUTURE SCOPE

The projected conclusion is an indigenous technique for developing an image processing-based intelligent system to monitor and regulate the solar panel's tracking ability. Once optimal output is achieved, this system is used in solar aerators to provide oxygen supply for fish farming in solar ponds. This technique uses a robust classifier for estimating the efficiency of the solar panel as compared to other conventional schemes. According to the performance requirements, the suggested classifier was able to recall all of the images of the sun at all inclinations with precision. The classification efficiency of the proposed WOPO is 99% in tracking the maximum power. The tracking time of WOPO is highly improved, and it is 21.14% faster.

An intelligent feed forward control for tracking the sun's location is conceivable as a result of the image analysis. Furthermore, at maximal intensities, the load generation is steady. As a result, a correlation between the intensity value of the sun's image and the expected power generation can be discovered. The major purpose of this research is to figure out what causes unfavourable power generation circumstances on foggy and rainy days. To maximize productivity, this automated technique assesses the sun's position using a set of quantitative parameters. This chapter discussed a design optimization of a solar-powered aeration system that includes a geo-spatial information system and intelligent image processing hybrid ANN algorithms. Using the proposed technique, a small-scale fish pond with a solar-powered aeration system meets the feasibility demand.

The future scope of this work would be to develop an Arduino-based hardware set-up for sun tracking system. This hardware would be capable of tracking the sun's position based on latitude and longitude position obtained from GPS corresponding to the solar images captured by the camera. These images will be processed by the various image processing and AI algorithms to estimate the power output.

REFERENCES

[1] Yoo, J., Kang, Y. (2014). Solar tracking system experimental verification based on GPS and vision sensor fusion. *Journal of Automation and Control Engineering*, 2, 4, 417–421.

[2] Badran, O., Arafat, I. (2013). The enhancement of solar distillation using image processing and neural network sun tracking system. *International Journal of Mining, Metallurgy & Mechanical Engineering*, 1, 3, 208–212.

[3] Lee, C.-D., Huang, H.-C., Yeh, H.-Y. (2013). The development of sun-tracking system using image processing. *Sensors*, 13, 5448–5459.

[4] Ruelas, A., Velázquez, N., González, L. (2013). Design, implementation and evaluation of a solar tracking system based on a video processing sensor. *International Journal of Advanced Research in Computer Science and Software Engineering*, 3, 10.

[5] Arturo, M. M., Alejandro, G. P. (2010). High–precision solar tracking system. *Proceedings of the World Congress on Engineering*, Vol: II.

[6] Sujatha, K. (2015). Wind turbine monitoring and control using fuzzy logic. *International Journal of Applied Engineering and Research*, 10, 11, 29231–29257.

[7] Ponmagal, R. S., Sujatha, K., Godhavari, T., Jegadeeswari, S. (2015). Big data analytics architecture for secured wireless body area sensor networks. *International Journal of Applied Engineering and Research*, 10, 68, 160–166.

[8] Sujatha, K., Ponmagal, R. S., Rajaram, U. (2014). Sensor cloud for autonomous navigation of unmanned ground vehicle. *International Journal of Engineering and Technology*, 2, 1, 1121–1126.

[9] Sujatha, K., Kumaresan, M., Ponmagal, R. S., Vidhushini, P. (2015). Vision based automation for flame image analysis in power station boilers. *Australian Journal of Basic and Applied Sciences*, 9 2, 40–45.

[10] Sujatha, K. (2014). Soft sensor for temperature measurement in gas turbine. *International Journal of Applied Engineering and Research*, 9, 23, 21305–21316.

[11] Sujatha, K. (2014). Brain computer interface for vehicle automation. *International Journal of Applied Engineering and Research*, 9, 24, 29403–29418.

[12] Sujatha, K. (2014). Combustion quality estimation in power station boilers using SVM based feature reduction with Bayesian. *European Journal of Scientific Research*, 120, 2, 189–198.

[13] Sujatha, K., Pappa, N., Senthil, K. K., Siddharth Nambi, U. (2013). Monitoring power station boilers using ANN and image processing, Trans Tech Publications, Switzerland, *Advanced Materials Research*, 631–632, 1154–1159.

[14] Sujatha, K., Pappa, N., Senthil, K., Siddharth Nambi, U., Raja Dinakaran, C. R. (2013). Intelligent parallel networks for combustion quality monitoring in power station boilers, Trans Tech Publications, Switzerland. *Advanced Materials Research*, 699, 893–899.

[15] Sujatha, K., Pappa, N., Senthil, K., Siddharth Nambi, U., Raja Dinakaran, C. R. (2013). Automation of combustion monitoring in boilers using discriminant radial basis network. *International Journal of Artificial Intelligence and Soft Computing*, 3, 3.

[16] Sujatha, K. (2012). Flame Monitoring in power station boilers using image processing. *ICTACT Journal on Image and Video Processing*, Dr. M.G.R Educational & Research Institute, Indexed in IET Inspec.

[17] Sujatha, K., Pappa, N. (2011). *Combustion Quality Monitoring in PS Boilers Using Discriminant RBF, ISA Transactions*, Elsevier, Vol: 2 (7), pp. 2623–2631.

[18] Sujatha, K., Ponmagal, R. S., Rajaram, U. (2014). Sensor cloud for autonomous navigation of unmanned ground vehicle. *International Conference on Defence and Security Technology*, pp. 97–100.

[19] Sujatha, K., Godhavari, T. (2015). Football match statistics prediction using artificial neural networks. *International Conference on Recent Trends in Information, Telecommunication and Computing*, pp. 189–198.

[20] Sujatha, K., Godhavari, T., Senthil Kumar, K., Malathi, M., Sinthia, P. (2015). Brain tumor detection and diagnosis using biomedical sensors and secure hybrid cloud. *International Symposium on Green Manufacturing and Applications*, 37–41.

[21] Sujatha, K., Godhavari, T., Senthil Kumar, K. (2015). An effective distributed service model for image based combustion quality monitoring and estimation in power station boilers. *International Conference on Computer and Communication*, 33–49.

[22] Sujatha, K., Siddharth Nambi, U. (2012). Intelligent flue gas monitoring in power station boilers. *Third International Conference on Theoretical and Mathematical Foundations of Computer Science*, 631–632.

[23] Sujatha, K., Siddharth Nambi, U. (2012). Monitoring power station boilers using artificial neural networks. *Third International Conference on Theoretical and Mathematical Foundations of Computer Science*, 633–635.

[24] Sujatha, K., Pappa, N. (2011). Combustion quality monitoring in power station boilers using SVM based feature reduction and RBF. *TIMA-2011 Trends in Industrial Measurements and Automatic*, 529–534, CSIR, Taramani.

[25] Sujatha, K. (2011). Flame image analysis for combustion quality estimation and power station boilers using classification algorithms. *Sustainable Energy and Intelligent System SEISCON'11*, Dr. M.G.R. Educational and Research Institute, 1134–1139.

[26] Sujatha, K., Pappa, N. (2010). *Combustion Quality Monitoring in Power Station Boilers, SYMOPA, CDAC*, 98–103.

[27] Sujatha, K., Venmathi, M., SheebaPercis, E., Nalini, A. (2010). *Detection of Flow in Rolling of Steel Sheets Using Image Processing, National Conference NCEEE10*, Sathyabama University, 129–133.

[28] Sujatha, K., Pappa, N. (2010). *Intelligent Sensor for Temperature Measurement in Power Station Boilers, National Conference on Intelligent Information Retrieval*, PSG College of Technology Coimbatore, NCIIR-2010, Vol: 2, pp. 1989–1994.

[29] Intelligent sensor for combustion quality and temperature measurement in power station boilers. In *Advanced Computing Sciences International Journal of Computational Intelligence*, Dr. Sujatha (Ed.), 2010, pp. 42–46.

[30] Science Analyst Limited Partnership, *Dissolved Oxygen Measurement.* Information on http://www.scienceanalys.com, 2016.

9 IoT-Based Dam and Barrage Monitoring System

Krishna Kumar
Indian Institute of Technology Roorkee

Gaurav Saini
Harcourt Butler Technical University

Rachna Shah
National Informatics Centre

Narendra Kumar
DIT University

Manoj Gupta
JECRC University

CONTENTS

9.1 INTRODUCTION

In June 2013, a mid-day cloudburst centered on the North Indian state of Uttarakhand caused devastating floods and landslides. Compared to other flood disasters in the nation, the length of the event was short. An extreme erosion took place because of the flash floods; the river banks were destroyed along the Kedarnath valley in the upstream portion of Kedarnath, besides the breach of Chorabari Lake and the deposition of debris/sediments in the valley. The disaster occurred due to the integrated impact of high intensity of rainfall, sudden breach of Chorabari Lake, and very steep

DOI: 10.1201/9781003272717-9

topography. Excessive discharge and silt impacted the hydropower generation and damaged the civil structures. The failure of the dam may also create a secondary disaster in the downstream areas. Therefore, monitoring dam/barrages is important. The dam breach parameters and peak flow prediction using the Froehlich method can be utilized to better design dams and barrages.

$$\text{Average breach width } (B_w) = 0.1803\,(V_w)^{0.32}\,(h_b)^{0.19} \tag{9.1}$$

$$\text{Failure time } (T_f) = 0.00254\,(V_w)^{0.53}\,(h_b)^{-0.9} \tag{9.2}$$

$$\text{Peak flow } (Q_p) = 0.607\,(V_w)^{0.295}\,(h_w)^{1.24} \tag{9.3}$$

where V_w is the volume of water stored above the breach at the time of failure (m^3), h_b is the height of the breach (m), and h_w is the depth of water above the breach at the time of failure (m). The 100-year (1901–2002) rainfall variation in Uttarakhand is shown in Figure 9.1, and the major water resources available in Uttarakhand are listed in Table 9.1.

9.1.1 Dam

A dam is a large physical structure that is constructed across a river or water source, whereas a barrage is a type of dam. Both dams and barrages cause a lake that can be used for various applications to develop hydropower, control flood, and provide irrigation. Different channels transport the water from these reservoirs,

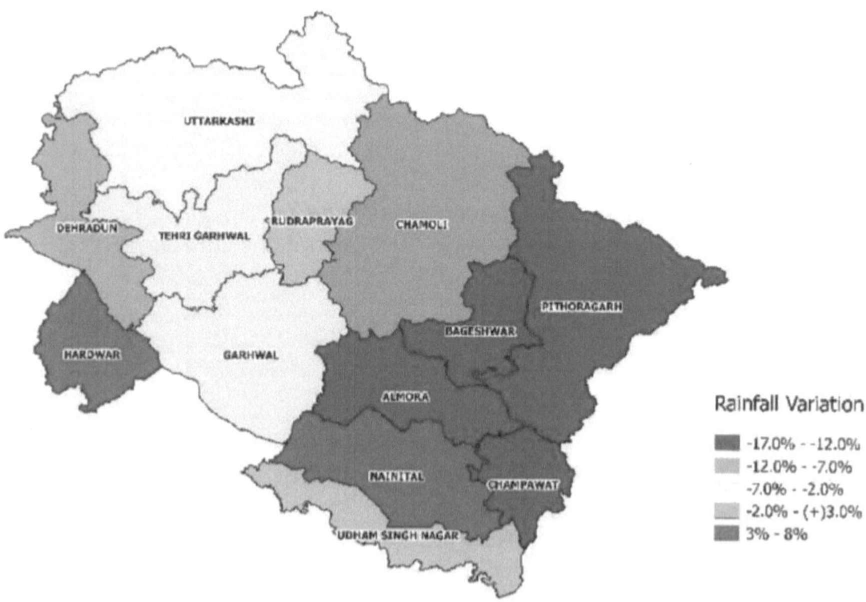

FIGURE 9.1 An analysis of 100 years (1901–2002) of rainfall data. (IMD rainfall data.)

TABLE 9.1
Major Water Resources of Uttarakhand

S. No.	District	Major Rivers/Tributaries
1	Almora	Kosi, Ramganga, Suyal, Gagas
2	Bageshwar	Saryu, Gomti, and Pindar
3	Chamoli	Alaknanda, Ramganga, Dhauliganga, Nandakini, Pindar
4	Champawat	Saryu, Kali/Sarada
5	Dehradun	Asan, Song, Tons, Rispina
6	Haridwar	Ganga, Solani
7	Nainital	Kosi, Gola, Nandhaur, Dabka, Baur, and Bhakra
8	Pauri Garhwal	Ganga, Alakhnanda, Nayar
9	Pithoragarh	Goriganga, Kali River, Saryu, Ramganga, Yangti, Dhauli, and Kuti
10	Rudraprayag	Mandakini, Alaknanda
11	Tehri Garhwal	Bhagirathi, Bhilangana, Alaknanda, Ganga
12	Udham Singh Nagar	Sarada, Gola, Phikka
13	Uttarkashi	Bhagirathi, Yamuna, Tons

FIGURE 9.2 Bhimgoda Barrage across the river Ganga.

which is thus used for vast applications. The dam has a massive structure that captures or stores the water during the monsoon period and releases it during the lean discharge period. However, the habitat may be affected by the construction of these structures. The barrages are small structures that divert the water flow for applications such as canal inlets and river hydroelectric power plants. The water behind the barrage creates a small storage to make the continuous supply of water and control the water flow entering the water channel. The Bhimgoda Barrage built across the river Ganga is shown in Figure 9.2, and the list of the top five countries having the maximum number of dams in the world is shown in Table 9.2.

TABLE 9.2
Top Five Countries with Number of Dams

S. No.	Country	Dams
1.	China	23842
2.	The USA	9261
3.	India	5102
4.	Japan	3112
5.	Brazil	1411

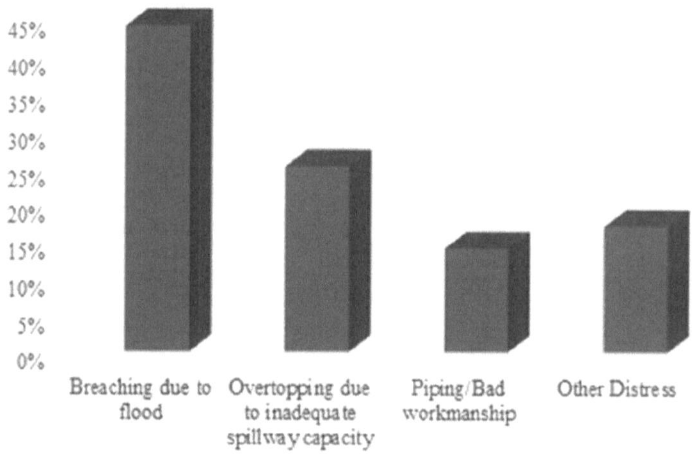

FIGURE 9.3 Reported dam failures.

Dams and barrages are usually constructed to control the water flow downstream or in the power channel. Various measures are required to control the water flow, which depends on different parameters, viz. water depth, discharge, silt concentration, climate conditions, structural behavior, and equipment conditions. To safeguard the nearby community and different areas, it is necessary to keep control over these parameters. The advancements in technologies ease the monitoring of dams and barrages, which can be used to effectively deal with the complex situations that arise during the operation and control of water flow. Various intelligent technologies, viz. IoT and artificial intelligence, can be used to monitor the parameters to control the water flow, especially during the monsoon period. Figure 9.3 shows the dam failure cases reported in India.

Figures 9.4 and 9.5 show the year-wise dam failure and state-wise dam failure scenarios, respectively. From year 1951 to 1960, a maximum of ten dam failures were reported. The maximum number of dam failures occurred in Madhya Pradesh and Rajasthan. Now, due to the development of advanced monitoring technologies, the failure of the dam has substantially been reduced.

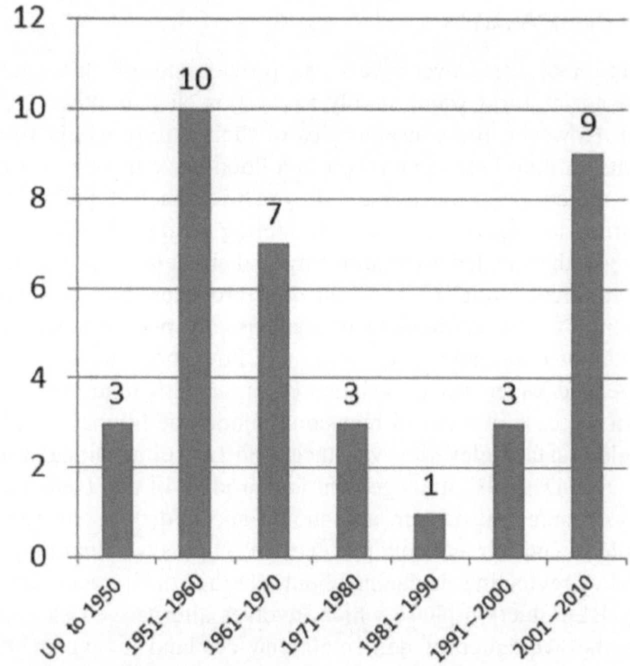

FIGURE 9.4 Year-wise dam failures in India.

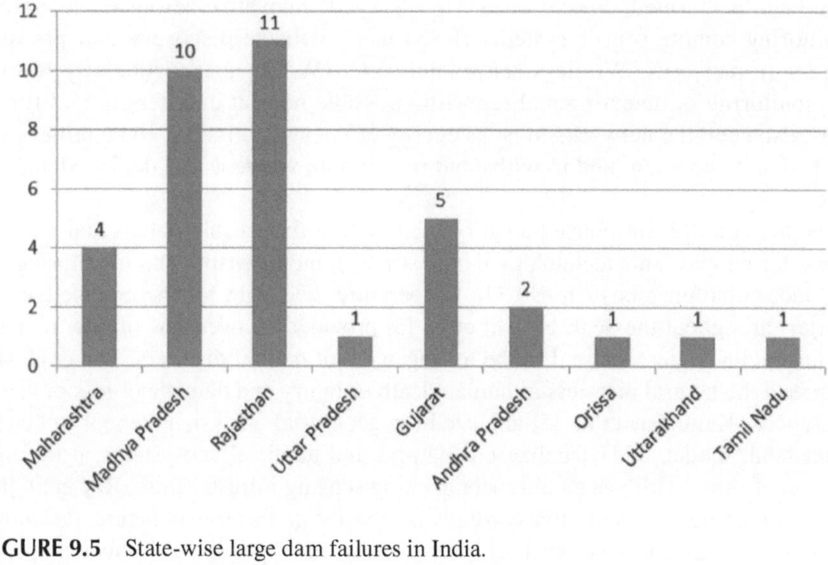

FIGURE 9.5 State-wise large dam failures in India.

9.1.2 Dam Break Analysis

The construction of dams over rivers can provide significant benefits such as drinking and agricultural water supply as well as electric power generation or flood control. However, the consequences of their failure would result in catastrophic events. A dam break can result in a flood wave traveling at high speeds up to tens of meters deep through a valley. The impact of such a wave can be very devastating for developed areas. Modeling of dam breaks can generally be done using either scaled hydraulic physical models or mathematical simulation. Mathematical modeling of flood dam breaches may be done by either one-dimensional or two-dimensional analysis. In one-dimensional analysis, information can be obtained in the direction of flow about the extent of the flood, i.e., discharge and water rates, variance of these with time, and flow rate via the breach. In the case of a two-dimensional model, additional details about the inundated field, surface elevation variance, and two-dimensional velocities can also be expected. Dam risk management is a product of two basic factors; i.e., a dam is always a potential danger, and should an incident occur, those involved need to be guaranteed fair and equal protection. The risk evaluation phase, which is the method of reviewing decisions about whether the current danger is tolerable and the risk reduction phase which involves alternative risk control strategies such as the avoidance of dam maintenance, land management, and civil safety steps to be taken along the valley preparedness. The primary aim of dam break risk reduction and hazard mitigation is to mitigate by both structural and non-structural interventions.

9.2 INVESTIGATIONS ON DAM AND BARRAGE MONITORING

To understand the applications of intelligent technologies, various investigations have been performed. Bindal et al. [1] surveyed innovative responses to fiasco monitoring remote sensor systems (RSS) using debacle discovery and precautionary frameworks. Wireless sensor networks (WSNs) are particularly useful for monitoring or detecting and reporting possible natural disasters in real time. They also need the networks to be as energy efficient as possible, to be reliable in disclosing its position, and to withstand the climate where it was deployed and to have a long life.

Pallavi et al. [2] introduced an IoT-based system that enables smart and remote access for people. This technology demonstrates remote sensing and monitoring of greenhouse parameters such as CO_2, temperature, and light for the complete crop quarter throughout the year. Eguchi et al. [3] provided an overview of how remote sensing technologies are used in the management of natural disasters. Van et al. [4] discussed the natural disasters in human death or injury and damage or loss of valuable goods. Kamilaris et al. [5] analyzed the geospatial analysis potential to better understand, model, and visualize our natural and artificial ecosystems, using the Internet of things (IoT) as an all-encompassing sensing infrastructure. Ray et al. [6] studied the disastrous events that cordially involve the momentum of nature. Recently, due to its attractive features such as heterogeneity, interoperability, lightweight, and flexibility, the IoT paradigm has opened a promising door to cater to a multitude of issues related to agriculture, industry, security, and medicine. A comparison between different IoT protocols is presented in Table 9.3 to understand the advantages and limitations.

Various researchers have also worked on the area of flood control, and a comparison between IoT-based disaster management systems proposed by researchers is presented in Table 9.4.

Shah et al. [13] studied the recent developments in big data analytics and IoT technologies for disaster management and to provide accurate insights. The role of disaster management system (DMS) applications and requirements is shown in Table 9.5.

A comparison among business development and analysis (BDAA) tools is shown in Table 9.6.

Sinha et al. [18] studied disaster management intending to mitigate potential disaster damage, ensure immediate and appropriate assistance to victims, and achieve effective and rapid recovery.

TABLE 9.3
Comparison between IoT Protocols

Parameters	Bluetooth	ZigBee	LoRa-WANs	WiMAX	NB-IoT
Standard	IEEE 802.15.1	IEEE 802.1.5.4	IEEE 802.15.4	IEEE 802.16	3GPP
Frequency range	2.4 GHz	2.4 GHz	868/915 MHz, 2.4 GHz	2–66 GHz	180 kHz
Transmission range	8–10 m	10–100 m	10–20 km	50 km	10–15 km
Cost	Low	High	Low	High	Medium

TABLE 9.4
Comparison between IoT-Based Disaster Management Systems

Article	IoT Architecture	Cloud-Enabled?	Features	Application Area
Ancona et al. [7]	Yes	No	Integrated intelligent WSN with IoT	Flood
Kumar et al. [8]	No	No	Machine-to-machine and ultra-low-power processing	Flood
Hernández-Nolasco et al. [9]	No	No	Integrated Wi-Fi with the developed system	Flood
Shalani et al. [10]	Yes	Yes	Messaging system	Flood
Lo et al. [11]	Yes	Yes	Image processing-based real-time flood monitoring	Flood
Inoue et al. [12]	Yes	Yes	Topology-independent data transfer facility	Landslide

TABLE 9.5
DMS Applications and Requirements

Disaster Status	DMS Application	DMS Requirements
Pre-disaster	Disaster prediction, early warning systems, and simulation exercises	Reliability, availability, maintainability, accuracy, and usability
Post-disaster	Evacuation, rescue assistance, monitoring/surveillance, logistics management	

TABLE 9.6
BDAA-Based Studies for Disaster Management

Reference	BDAA Tools	Data Source	Text Analysis	Spatial Analysis	Focus
De et al. [14]	Spark	Crowdsourced sensor data	Yes	Yes	Real-time flood monitoring
Lin et al. [15]	Spark	Historical data from the metrological center	Yes	Yes	Simulation for typhoon risk assessment
Asencio-Cortés et al. [16]	Spark	Historical database of earthquake catalogs	Yes	Yes	Earthquake magnitude prediction
Wang et al. [17]	Hadoop	Social media, remote sensing	Yes	Yes	Coordination during disaster

From the literature review, it has been found that various techniques and devices are available to monitor and control dams and barrages, but due to the high cost of automation tools, monitoring small dams and barrages is neglected.

9.3 CIRCUIT CONFIGURATION FOR MONITORING AND CONTROL OF DAMS/BARRAGES

NodeMCU is an open-source firmware and development kit; it includes firmware that runs on the ESP8266 Wi-Fi microcontroller unit (MCU), which means it is a computer on a single chip. A piezoelectric sensor is used for the measurement of dam vibration. Barometric pressure sensors measure the absolute pressure of the air around. The water flow (discharge) sensor consists of a pinwheel sensor that measures the quantity of water that has passed through it. A float switch is used to sense the dam water level. A relay module is connected to operate the gates of the dam and barrage remotely. Figure 9.6 shows the circuit configuration of the IoT-based dam and barrage monitoring system.

Sensors are connected to NodeMCU, and the NodeMCU module is connected to the cloud gateway. Various open-source web services are available. ThingSpeak is

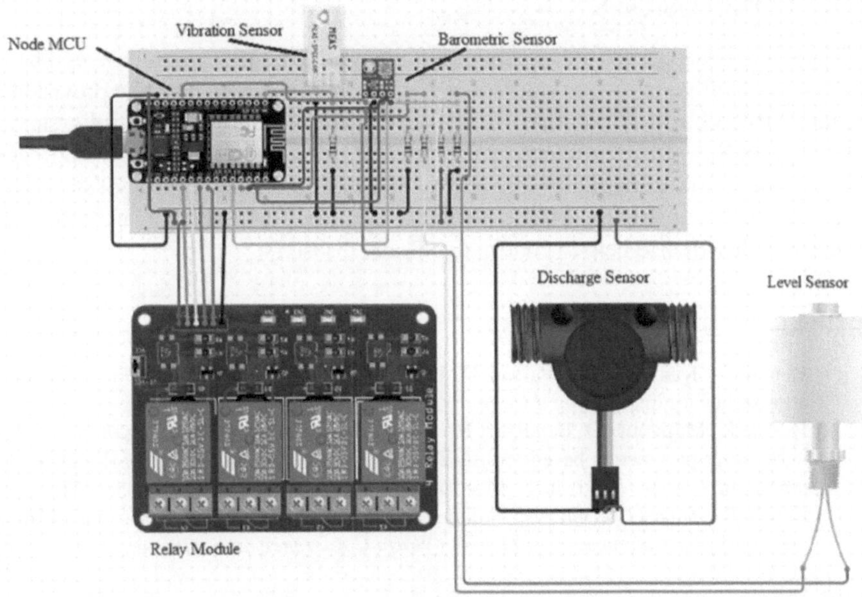

FIGURE 9.6 Circuit configuration for dam/barrage monitoring.

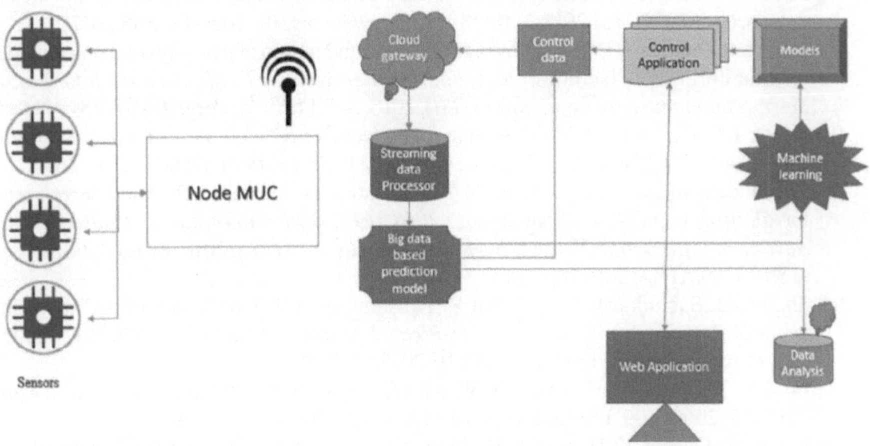

FIGURE 9.7 IoT architecture for dam/barrage monitoring and controlling system.

used in our case. ThingSpeak is an IoT framework and an open-source API to store and retrieve hardware devices and sensors data. It uses HTTP protocol for its communication over the Internet or LAN. It also has the facility to analyze and visualize hardware or sensor devices data. The creation of channels for each sensor data is simple. These channels can be set as private channels or public channels that can share the data publicly. Figure 9.7 shows the architecture of the proposed model.

9.4 CONCLUSIONS

The dam and barrage are the structures mainly used to control water flow, and based on the availability and requirement of water, it can be released with the provision of control. To tackle water control and reduce the risk of hazard, a control system is essential. In this paper, an Internet of things (IoT)-based system is presented to control the water flow through the dam and monitor the condition of the civil structures. The proposed architecture will be helpful to operate a dam and barrage safely and efficiently. It will also help to control floods and disasters.

REFERENCES

[1] Bindal A, Kadhim MH, Parsad D, Patel RB. A pragmatic review on algorithmic approaches for disaster. *Int J Comput Corp Res* 2014, 4.

[2] Pallavi K, Mallapur JD, Bendigeri KY. Remote sensing and controlling of greenhouse agriculture parameters based on IoT. *2017 International Conference on Big Data, IoT and Data Science (BID)*, 2018, 44–8. https://doi.org/10.1109/BID.2017.8336571.

[3] Eguchi RT, Huyck CK, Ghosh S, Adams BJ. The application of remote sensing technologies for disaster management. *The 14th World Conference on Earthquake Engineering*, 2008, 17 p.

[4] Van Westen C. Remote sensing for natural disaster management. *Int Arch Photogramm Remote Sens Spat Inf Sci ISPRS Arch* 2000, 33, 1609–17.

[5] Kamilaris A, Ostermann F. *Geospatial Analysis and Internet of Things in Environmental Informatics*, 2018, 1–8.

[6] Ray PP, Mukherjee M, Shu L. Internet of things for disaster management: State-of-the-art and prospects. *IEEE Access* 2017, 5, 18818–35. https://doi.org/10.1109/ACCESS.2017.2752174.

[7] Kim H, Shin J, Shin H, Song B. Design and implementation of gateways and sensor nodes for monitoring gas facilities. *2015 Fourth International Conference on Information Science and Industrial Applications (ISI)*, 2016, 3–5. https://doi.org/10.1109/ISI.2015.15.

[8] Ashok KV, Girish B, Rajesh KR. Integrated weather & flood alerting system. *Int Adv Res J Sci Eng Technol* 2015, 2, 21–4. https://doi.org/10.17148/IARJSET.2015.2606.

[9] Hernández-Nolasco JA, Ovando MAW, Acosta FD, Pancardo P. Water level meter for alerting population about floods. *2016 IEEE 30th International Conference on Advanced Information Networking and Applications (AINA)*, 2016, 879–84. https://doi.org/10.1109/AINA.2016.76.

[10] Shalini E, Subbulakshmi S, Surya P, Thirumurugan R. Cooperative flood detection using SMS through IoT. *Int J Adv Res Electr Electron Instrum Eng* 2007, 3297, 2099–103. https://doi.org/10.15662/IJAREEIE.2015.0503138.

[11] Lo SW, Wu JH, Lin FP, Hsu CH. Visual sensing for urban flood monitoring. *Sensors* 2015, 15, 20006–29. https://doi.org/10.3390/s150820006.

[12] Inoue M, Owada Y, Hamaguti K, Miura R. Nerve net: A regional-area network for resilient local information sharing and communications. *2014 Second International Symposium on Computing and Networking*, CANDAR 2014, 2015, 3–6. https://doi.org/10.1109/CANDAR.2014.83.

[13] Shah SA, Seker DZ, Hameed S, Draheim D. The rising role of big data analytics and IoT in disaster management: Recent advances, taxonomy, and prospects. *IEEE Access* 2019, 7, 54595–614. https://doi.org/10.1109/ACCESS.2019.2913340.

[14] de Assis LFFG, Horita FEA, de Freitas EP, Ueyama J, de Albuquerque JP. A service-oriented middleware for integrated management of crowdsourced and sensor data streams in disaster management. *Sensors* 2018, 18. https://doi.org/10.3390/s18061689.

[15] Lin S, Fang W, Wu X, Chen Y, Huang Z. A spark-based high performance computational approach for simulating typhoon wind fields. *IEEE Access* 2018, 6, 39072–85. https://doi.org/10.1109/ACCESS.2018.2850768.

[16] Asencio-Cortés G, Morales-Esteban A, Shang X, Martínez-Álvarez F. Earthquake prediction in California using regression algorithms and cloud-based big data infrastructure. *Comput Geosci* 2018, 115, 198–210. https://doi.org/10.1016/j.cageo.2017.10.011.

[17] Wang Z, Vo HT, Salehi M, Rusu LI, Reeves C, Phan A. A large-scale spatio-temporal data analytics system for wildfire risk management. *GeoRich'17: Proceedings of the Fourth International ACM Workshop on Managing and Mining Enriched Geo-Spatial Data*, Conjunction with SIGMOD 2017, 2017, 19–24. https://doi.org/10.1145/3080546.3080549.

[18] Sinha A, Kumar P, Rana NP, Islam R, Dwivedi YK. Impact of internet of things (IoT) in disaster management: A task-technology fit perspective. *Ann Oper Res* 2019, 283, 759–94. https://doi.org/10.1007/s10479-017-2658-1.

10 Complex Hydrides *Lightweight, High Gravimetric Hydrogen Storage Materials*

Vivek Shukla and Thakur Prasad Yadav
Banaras Hindu University

CONTENTS

10.1 INTRODUCTION

Since the birth of civilization, fossil fuels have fulfilled almost all of humanity's energy needs. Even if their availability is guaranteed till the end of the century, the negative impact of their use has come to the fore in recent years.[1] Earlier depletion of fossil fuels and urban air pollution were the primary concerns. At present, the

DOI: 10.1201/9781003272717-10

climate change effects produced by the accumulation of CO_2 (produced by burning fossil fuels) in the upper atmosphere have become the dominant deteriorating factor [1–5]. The CO_2 continues to increase at an alarming pace. Currently, the CO_2 concentration is approximately 414 ppm [6]. It must be reduced to 200 ppm (half the current level) by 2050 so that the change in ambient temperature does not exceed 2°C (Paris Convention on Climate Change 2015) [7]. The resulting climate change effect could make the earth habitable in several decades. Another problem is the large gap between energy demand and supply. This continues to grow as the population grows. There will be 10 billion of them by 2050. The Paris convention asserted that a substantial fraction of coal, oil, and gas reserves would have to remain underground and unburned (stranded assets) to ensure a cap on CO_2 emission (the main cause of climate change) so as to keep the temperature rise within 2°C limit [6,7]. Under these circumstances, we have to fall back more dominantly and perhaps solely on energy derived from clean, renewable, and indigenous energy systems.

Decades of research have revealed that, among several choices, hydrogen is the most appealing candidate [1–5,8–11]. Hydrogen, which can be produced by water and burned back to water through combustion with atmospheric oxygen (hot combustion in internal combustion (IC) engines and cold combustion in fuel cells). It is thus completely a renewable, non-exhaustible, environment-friendly, and indigenous fuel. Therefore, it has been observed that hydrogen is considered to be one of the most cost-effective renewable energy options out of the whole available energy use scenario. It is commonly known that hydrogen fuel emits no carbon dioxide and that the dominant combustion product is water, making hydrogen the ideal fuel for preventing climate change. Recently, there has been an increase in the research in the

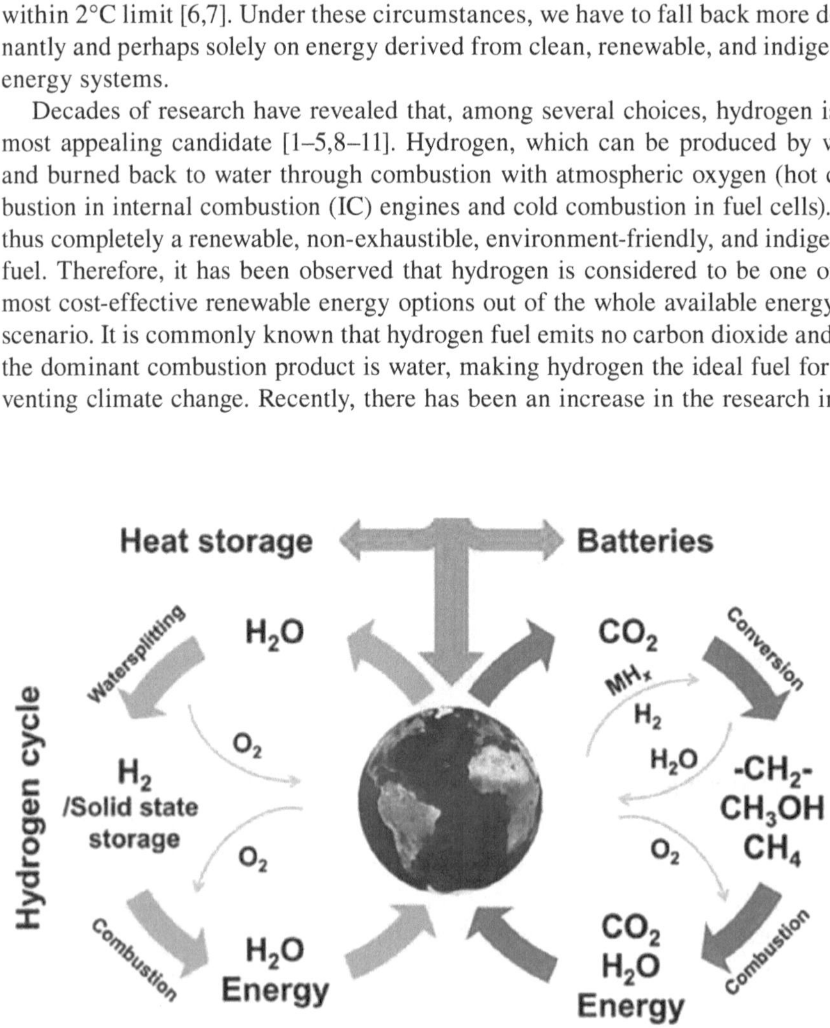

FIGURE 10.1 Schematic representation of hydrogen as a sustainable energy system [14].

field of synthesis, storage, and use of hydrogen. The most important aspect of this is warehousing, which includes production/distribution, safety, and service. After decades of research, it has been recognized that, compared to storing hydrogen in the form of compressed gas or liquid, storing hydrogen in the solid state as hydride is the safest and most efficient method of storage [12,13] (Figure 10.1).

The gaseous storage of hydrogen requires a very high pressure (>100 atm) to store it in cylinders; however, hydrogen storage in the form of a hydride tank does not require high pressure. Hydrogen storage in the metal hydride also doesn't require a low temperature (~252°C) as it is required for liquid storage. That is why hydrogen storage in the form of metal hydrides is a safe and efficient mode of storage. For the practical use of these metal hydrides, the US DOE has set some targets for materials to become viable hydrogen storage materials (the US DOE 2020 target is 4.50 wt.% by weight and 30 g/L, operating temperature ~50°C–100°C, life cycles ~1000 cycles), rapid sorption kinetics (~4 wt.% H_2/minute), controlled thermodynamics, and minimal loss of capacity during the cycle [15]. Storing hydrogen in metal hydrides is known to be the most efficient and safest storage method. Until now, no hydrogen storage material has met all the requirements. However, significant progress has been made, and research and development continue to address the challenges of hydrogen storage in hydrides [16,25] (Figure 10.2).

Nevertheless, the research focus has shifted from the last two decades to lightweight hydride as intermetallic hydrides are generally made of transition metal elements and are heavy. These are takeout and fill-in type hydrides such as metal alanates and metal borohydrides, and others. The advantage of this lightweight material is that it meets the US DOE (Department of Energy) criteria for onboard hydrogen storage. However, these materials' desorption temperatures are high (>350°C), their sorption (de-/rehydrogenation) kinetics are slow, and their cyclability is poor [18,19].

10.2 HYDROGEN STORAGE IN COMPLEX METAL HYDRIDES

Boron, nitrogen, and aluminum are some light elements that can construct complex metal hydrides. They have low thermodynamic and kinetic characteristics together with high hydrogen densities [17–19]. The discovery of the reversibility of titanium-catalyzed $NaAlH_4$ from hydrogen storage to complex anions for the first time triggered a paradigm shift in research [20]. Therefore, the research focused on metal borohydrides such as $LiBH_4$ and complex nitrogen-based hydrides such as $LiNH_2$ [21–27]. In this type of complex hydrides, the bond between anions is generally covalent with well-defined directionality.

In contrast, there is an ionic bond between complex anions and countercations in the solid state, e.g., $LiBH_4$, $NaBH_4$, and $NaAlH_4$. The large gravimetric (18.6 wt.%) and volumetric densities of these built-in hydrides make them attractive hydrogen storage materials. The hydrogen storage properties of various complex hydrides are shown in Table 10.1. However, the high thermodynamic stability and kinetic barrier only allow hydrogen release at a very high temperature. This restricts the materials from becoming viable hydrides [17–25]. Therefore, they are not suitable for practical use. Furthermore, since the pathway for hydrogen release from complex hydrides is more complicated than from metal and ionic hydrides, it is not yet fully understood.

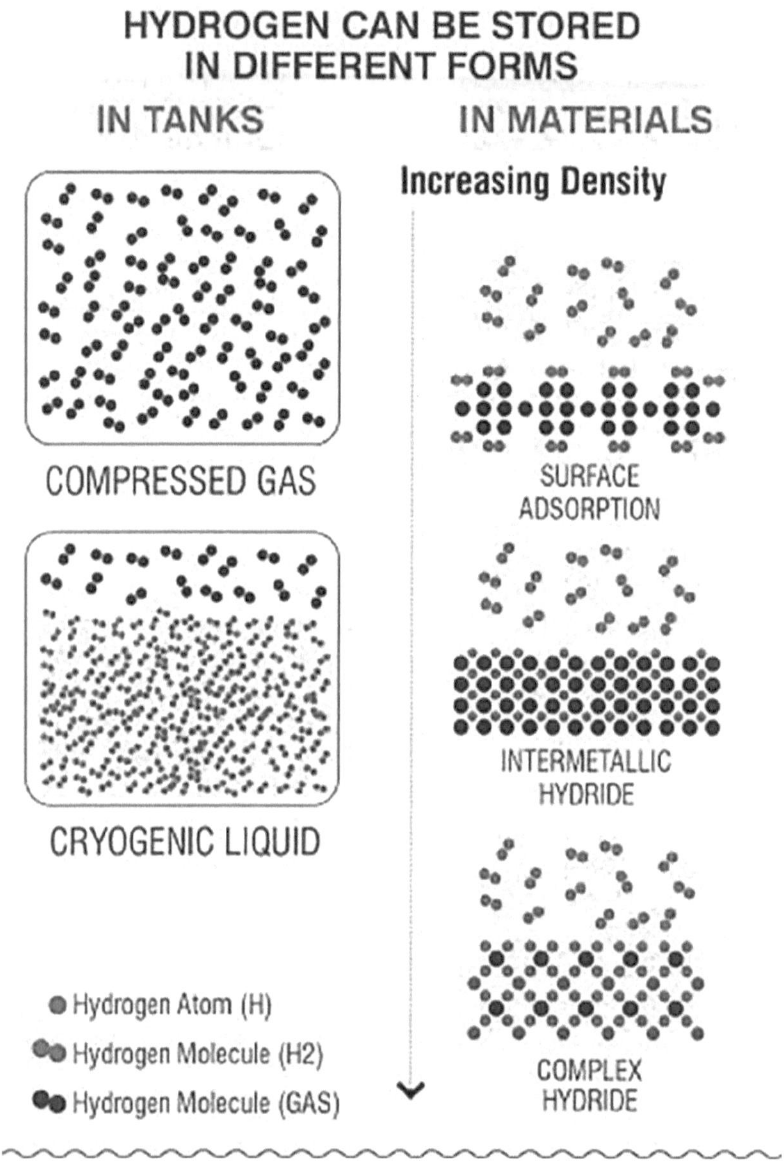

FIGURE 10.2 Storage of hydrogen in different modes [17].

10.2.1 Metal Borohydride

Recently, a lot of work has been done on the metal borohydrides systems. However, very high temperatures are required to separate hydrogen from the borohydrides. The various borohydrides' crystal structures can be found in a review [31]. One of the basic borohydrides is $LiBH_4$. The $LiBH_4$ decomposes at 438°C to LiH and B. This releases nearly 13.5 wt.% hydrogen according to the following equation:

$$2LiBH_4 \rightarrow 2LiH + 2B + H_2 \quad (10.1)$$

TABLE 10.1
The Essential Feature of Different Complex Hydrides [20–21]

	M (g/mol)	ρ (g/mL)	ρ_m (wt.% H_2)	ρv (gH_2/L)	ΔH_{dec} (kJ/mol)	T (1 bar) (°C)	T_{dec} (°C)
$LiBH_4$	21.78	0.66	18.4	122.5	74	370	~400
$NaBH_4$	37.83	1.07	10.8	115.6	108	534	~500
$LiAlH_4$	37.95	0.92	10.6	97.5	−10	-	~150[a]
Li_3AlH_6	53.85	1.02	11.2	114.2	25	−81[c]	~200[a]
$NaAlH_4$	54.00	1.28	7.3	93.4	33.1	18	~230[a]
Na_3AlH_6	102.00	1.45	5.9	85.6	49.0	103	~275[a]
$LiNH_2$	22.96	1.18	8.8	103.6	67[b]	-	~300

M, molar mass; T_{dec}, decomposition temperature; T, equilibrium temperature; ΔH_{dec}, decomposition enthalpy; ρ, volumetric mass density; ρ_v, volumetric hydrogen density; ρ_m, gravimetric hydrogen density.

[a] Not catalysed.

[b] Reported for the $LiNH_2$-LiH system.

[c] Calculated based on the van't Hoff equation using change in entropy = 130 J/(mol K).

Due to high thermodynamic stability, the research focus has been to destabilize these borohydrides. Different approaches have been adopted to destabilize these borohydrides, e.g., employing suitable catalysts, additives, and making nanocomposites.

Recently, it has been found that reducing the size of the material to the nanometer range is an effective way to reduce the de-/rehydrogenation kinetics of the materials. However, it has been observed that these nanomaterials during repeated de-/rehydrogenation get agglomerated. If we incorporate these nanocomposites' carbon nanostructure, the agglomeration can be avoided after repeated cycling [27–34]. Also recently, much attention has been paid to the alkaline earth borohydrides and the transition metals $Mg(BH_4)_2$, $Ca(BH_4)_2$, $Zr(BH_4)_4$, and $Zn(BH_4)_2$ [35,36]. These borohydrides are much less stable than Group 1 borohydrides (e.g., $LiBH_4$ and $NaBH_4$). These borohydrides seem very promising materials due to low thermodynamic stability as they desorb hydrogen at low temperatures. But the problem is their reversibility that exists as in the case of the less stable metal alanates. These materials have shown promising potential at moderate decomposition temperatures, but the problem of reversibility still exists, as is the case with less stable alanates. Recently [34], the synthesis of $Ca(BH_4)_2$ is directly formed by CaB_6, CaH_2, and H_2 gas at a pressure of 700 bar and a temperature of 400°C–440°C. However, this material doesn't show complete reversibility. Many studies have been done on this material, but the high thermodynamic stability and sluggish kinetics render its application for vehicular transportation difficult (Figure 10.3).

10.2.2 Metal Aluminum Alanates

The dehydrogenation complex metal aluminum hydride can be understood by the following stages. These three stages of desorption of metal alanates are implicit in the following equations [35].

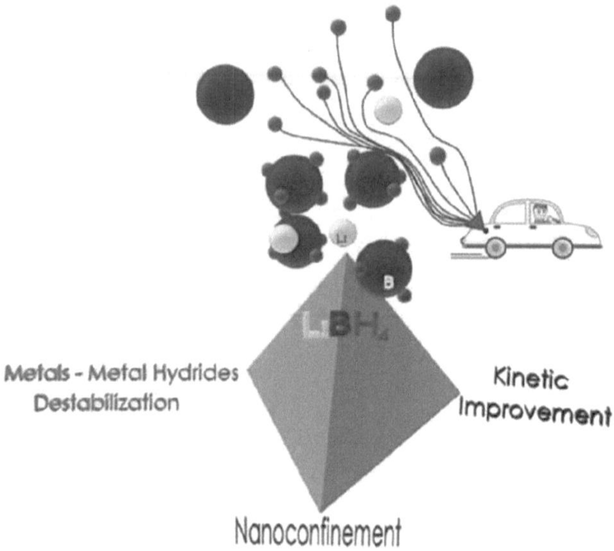

FIGURE 10.3 Destabilization of $LiBH_4$ and schematic diagram for understanding the mechanism [37].

$$3XAlH_4 \rightarrow X_3AlH_6 + 2Al + 3H_2 \tag{10.2}$$

$$X_3AlH_6 \rightarrow 3XH + Al + 3/2H_2 \tag{10.3}$$

$$3XH \rightarrow 3X + 3/2H_2 \tag{10.4}$$

Here, X denotes the alkali element. Step 3 involves the breakdown of hydride at high temperatures. As a result, when considering complicated aluminum hydride for hydrogen storage, step 3 is usually skipped. Many experiments on alanates have been conducted to enhance the thermodynamics of stages 1 and 2. Among the complicated aluminum hydrides, $NaAlH_4$, the prototype alanate, has received a lot of attention [38,39]. Due to its theoretically reversible hydrogenation power of 5.6 wt.%, low cost, and volume presence, sodium aluminum hydride ($NaAlH_4$) apparently is a profitable implementation option as a practical material in the side of hydrogen. The alanates [40–48] are the complicated hydride group that has garnered the most attention in the recent decade. Alanates are notable for their large storage capacities; nevertheless, they disintegrate in two phases when dehydrogenated.

$$3NaAlH_4 \rightarrow Na_3AlH_6 + 2Al + 3H_2 \ (3.7 \text{ wt.\%}) \tag{10.5}$$

$$Na_3AlH_6 \rightarrow 3NaH + Al + 1.5H_2 \left(1.8 \text{ wt.\%}\right) \tag{10.6}$$

The $NaAlH_4$ releases 3.7 wt.% of hydrogen in the first step as in equation 10.7 and is converted into Na_3AlH_6. The Na_3AlH_6 then desorbs 1.8 wt.% of hydrogen and produces the products NaH and Al.

The reversibility of these reactions is an important factor in making these materials viable hydrogen storage materials for practical applications. In spite of the very high hydrogen storage capacity, the irreversibility and slow kinetics of these hydrides restrict their practical applications. When titanium-doped $NaAlH_4$ (1997) demonstrated reversible dehydrogenation and rehydrogenation, it has been demonstrated reversible up to a certain extent. The breakthrough in these materials was made by employing Ti-based compounds (such as $TiCl_3$ or $Ti[OBu]_4$ to $NaAlH_4$) to lower the hydride's first breakdown temperature. When the temperature reaches 353 K, a quantity of 3.7 wt.% is dehydrogenated. However, the hydrogen content is reduced from 5.5 wt.% in the hydride without a catalyst [37]. Furthermore, the reaction is reversible; at 270°C at 175 bar hydrogen pressure, full conversion to the product was obtained in 2–3 hours [38]. The working temperatures for the first and second reactions are 185°C–230°C and 260°C, respectively. After that, the NaH started to break at higher temperatures with a total release of hydrogen of 7.4 wt.%.

As discussed, NaH only destabilized at the very higher temperature of 425°C, which is practically impossible. However, the first two processes must be considered for hydrogen storage [49]. After 17 cycles [50], using 2 mol.% TiN as a doping agent, a cyclic storage capacity of 5 wt.% H_2 is reached. However, the hydrogenation rate decreases with the number of cycles [49]. This clearly shows that titanium improves the de-/rehydrogenation kinetics reaction of $NaAlH_4$ [40–44]. The reversible hydrogen storage capacity on the cycling of sodium alanate catalyzed Ti-catalyzed over 100 cycles with the hydrogen storage capacity of 4 wt.% at 160°C. The hydrogen desorption enthalpy of $NaAlH_4$ doped with Ti to Na_3AlH_6 and Al had been found to be 37 kJ/mol. The enthalpy change calculated for this study for dehydrogenation is in agreement with the previous studies. Based on this value, the temperature required for the hydrogen pressure to equilibrate of 0.1 MPa was determined to be 33°C.

Conversely, the possible values of temperature for hydrogen pressures of 0.2 and 0.7 MPa are 60°C and 80°C, respectively. Unfortunately, in the temperature range of 70°C–100°C, the equilibrium hydrogen pressure for the Na_3AlH_6/NaH Al + H_2 equilibrium is inadequate. Therefore, it is not sufficient for use in PEM fuel cell systems where the heat of the exhaust vapor is used to generate ingredients.

In the case of other alanates, which can hydrogenate under moderate temperature and pressure, sorption is Na_2LiAlH_6, $KAlH_4$, K_3AlH_6, K_2LiAlH_6, and K_2NaAlH_6. In these alanates, the restriction is lower gravimetric densities compared to $NaAlH_4$. The most stable alanates exist in $NaAlH_4$. On the other hand, some alanates such as $LiAlH_4$, Li_3AlH_6, $Mg(AlH_4)_2$, $Ca(AlH_4)_2$, and $Ti(AlH_4)_4$ are known to be less stable than sodium alanate with high hydrogen storage capacity. Moreover, these materials are irreversible under moderate conditions (temperature and pressure) despite their promising hydrogen sorption characteristics (Figure 10.4).

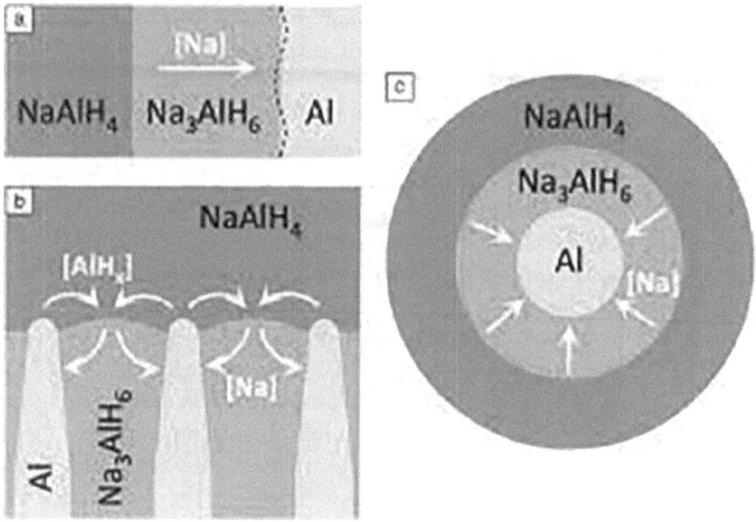

FIGURE 10.4 Distinct types of morphologies for $NaAlH_4$ [47].

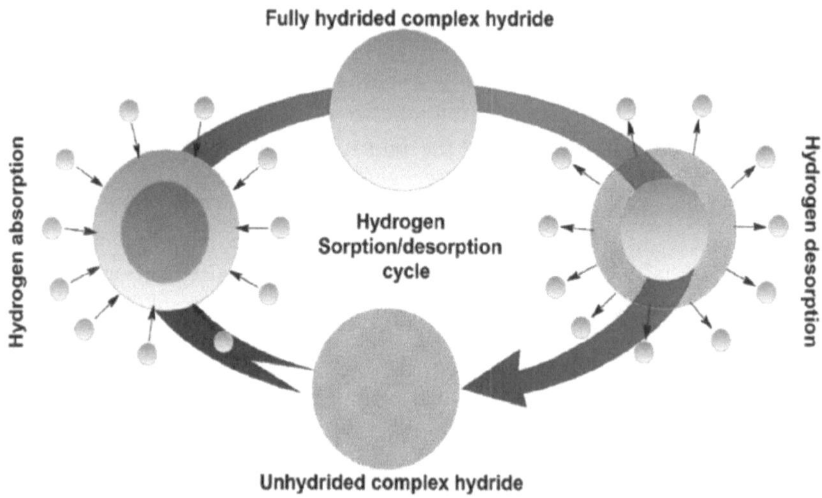

FIGURE 10.5 De-/rehydrogenation in amide/hydride composites [53].

10.2.3 Amide/Imides for Hydrogen Storage Applications

10.2.3.1 Potential Hydrogen Storage Material: Li-N-H System

Chen et al. (2002) carried out a significant work on the Li-N-H system. In 2003, Chen et al. studied the reversible hydrogen sorption in the lithium nitride (Li_3N) [51–53]. Light-element hydrogen storage materials are ideal for onboard hydrogen storage. These materials can afford a nearly 11.5 wt.% gravimetric density. Thus, this material shows excellent hydrogen good hydrogen storage capacities. Hydrogenation of

Li_3N occurs in two stages. In the first step, an imide hydride product is produced, and in the second, an amide hydride has been found [45,46].

$$Li_3N + H_2 \leftrightarrow Li_2NH + LiH \Delta H = -116\,kJ/mol \tag{10.7}$$

$$Li_2NH + H_2 \leftrightarrow LiNH_2 + LiH \Delta H = -45\,kJ/mol \tag{10.8}$$

The H_2 storage capacity (theoretical) of the above-discussed two processes is 5.5 and 6.5 wt.%, respectively. Li_2NH is produced via partial hydrogenation of Li_3N. Li_3N, when absorbed by hydrogen, produces LiH. At the same time, H is replaced by Li, which then combines with N to form Li_2NH (Figure 10.5). In the amide hydride reaction, $LiNH_2$ combines with LiH to produce Li_2NH, as shown by the reverse reaction in equation 10.8.

10.2.3.2 Li-Mg-N-H System

To destabilize the Li-N-H system, partial replacement of Li by Mg had been done. This significantly reduces the desorption enthalpy to 40 kJ/mol. Several groups [54–61] reported the interaction between $LiNH_2$ and MgH_2 (or $Mg(NH_2)_2$ and LiH). The optimum ratio for $Mg(NH_2)_2$-LiH has been found at 1:2. Several parameters in the reaction were examined, including various amide/hydride ratios and de-/rehydrogenation settings.

$$2LiNH_2 + MgH_2 \rightarrow Li_2Mg(NH)_2 + 2H_2 \leftrightarrow Mg(NH_2)_2 + 2LiH \tag{10.9}$$

Luo et al. (2006) reported comparing the re-/dehydrogenation stages of the 2:1 $LiNH_2+MgH_2$ and 1:1 $LiNH_2+LiH$ systems. The Mg-substituted Li-N-H system (Li-Mg-N-H system) started to desorb the hydrogen at a lower temperature (100°C) compared to the Li-N-H system. The P-C-T curves indicate significant differences: The hydrogen equilibrium for the Li-Mg-N-H composite has a pressure of 50 bar at 220°C, and for the Li-N-H combination, 1 bar of H_2 at 280°C. Similar to the $LiNH_2$-LiH reaction, an amide hydride reaction with $Ca(NH_2)_2$ and CaH_2 was observed. CaNH was formed when heated to 300°C, and Ca_2NH was formed when heated to 500°C [59]. The synthesis of $Ca(NH_2)_2$ can be done by mechanochemical ball milling of calcium hydride (CaH_2) in an NH_3 atmosphere at ambient temperature [59]. To make it a viable storage material, the working temperature of Li-Ca-N-H materials should be lowered. More research is needed to realize and optimize the system's hydrogen storage performance.

10.2.3.3 Destabilization of Li-Mg-N-H through the Addition of Metal Borohydrides

To get viable hydrogen storage materials, a lot of attention has been paid to the destabilization of the Li-Mg-N-H system. The kinetics and thermodynamics of this system can be improved by adding metal borohydrides. The binary systems $2LiBH_4+MgH_2$ [60], $2LiNH_2+LiBH_4$ [61], and $2LiNH_2+MgH_2$ [62–64] show rich chemistry. One crucial investigation on the Li-Mg-B-N-H system was done by Yang et al. in 2007. In their work, they have investigated the hydrogen sorption of the

2:1:1 $LiNH_2+MgH_2+LiBH_4$ composite. In their study [65], combinatorial synthesis and screening techniques revealed that the optimal composition of this three-component system was 6:3:1 $LiNH_2+MgH_2+LiBH_4$. For this ternary composition, a self-catalyzed reaction mechanism had been proposed: (i) $LiNH_2$ reacts with $LiBH_4$ in the molar ratio of 3:1 to form $Li_4BN_3H_{10}$ during ball milling, following equation 10.10. (ii) After that, $Li_4BN_3H_{10}$ reacts with MgH_2 producing $Li_2Mg(NH)_2$ (as shown in equation 10.11). (iii) Finally, the produced $Li_2Mg(NH)_2$ acts as a seed for the reaction described in equation 10.9. This results in improved hydrogen storage properties of the LiMgNH material [65]. The predominant seeding effect in improving the dehydrogenation kinetics of equation 10.10 was confirmed by a similar group [64]. A disadvantage of the ternary system is that the reversible H_2 capacity is slightly smaller than that of the original system. The ideal mixture has a maximum reversible H_2 level of only 3.5 wt.%:

$$3LiNH_2+LiBH_4 \rightarrow Li_4(BH_4)(NH_2)_3 \tag{10.10}$$

$$2Li_4(BH_4)(NH_2)_3+3MgH_2 \rightarrow 3Li_2Mg(NH)_2+2LiBH_4+6H_2 \tag{10.11}$$

It is expected that during the dehydrogenation reaction, the $LiNH_2$ and MgH_2 may be converted to $Mg(NH_2)_2$ and LiH at a certain condition (temperature and pressure), following equation 10.9. The thermal effects and chemical adaptations of 1:2:1 $Mg(NH_2)_2+LiH+LiBH_4$ and 2:1:1 $LiNH_2+MgH_2+LiBH_4$ were investigated. By varying the composition of LiBH4, Hu et al. (2008) compared the hydrogen sorption performance of these systems in detail. Hu et al. (2008) confirmed the conversion of MgH_2 and $LiNH_2$ into $Mg(NH_2)_2$ and LiH in the presence of $LiBH_4$ at 120°. By varying the MgH_2 ratios within the $2LiNH_2+LiBH_4+x(MgH_2)$, similar results

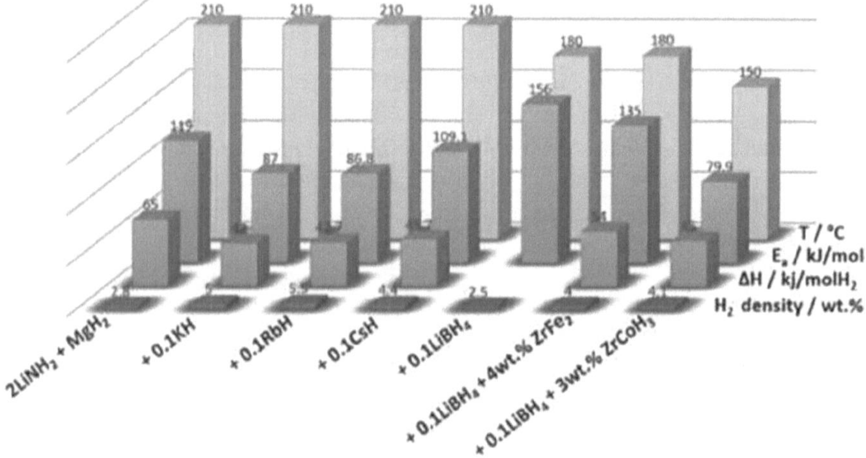

FIGURE 10.6 Comparison of kinetic and thermodynamic data for the catalyzed and uncatalyzed 2:1 $LiNH_2$-MgH_2 [53].

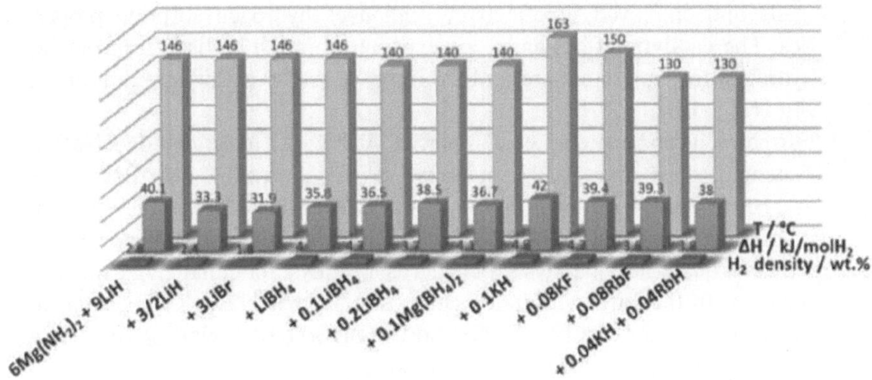

FIGURE 10.7 Comparison data of kinetics and thermodynamics of catalyzed and uncatalyzed 1:2 $Mg(NH_2)_2$-LiH [53].

have been found. MgH_2 reacts with $LiNH_2$ to form $Mg(NH_2)_2$ and LiH together with $Li_4(BH_4)(NH_2)_3$ during ball milling, as shown in equation 10.11. The comparison data of kinetics and thermodynamics of the Li-Mg-N-H system are shown in Figures 10.6 and 10.7.

10.2.3.4 Different Approaches for Improving the Hydrogen Sorption Performance of 1:2 $Mg(NH_2)_2$-LiH by the Employment of a Suitable Catalyst

In order to destabilize and form a pure basic phase of 1:2 $Mg(NH_2)_2$-LiH, the following approach should be taken.

a. Use of a suitable catalyst that is harder than the starting $LiNH_2$ and MgH_2. The synthesis of the pure phase of the basic system 1:2 $Mg(NH_2)_2$-LiH has proved to be difficult. This issue has been solved in the research carried out by using 2:1 $LiNH_2$-MgH_2 together with the catalyst $ZrFe_2$, which serves the twin role of catalyst and pulverizer. These components have been mixed and turned into nanoparticle form through ball milling. In this way, a pure phase is formed, and the hydrogen sorption properties are also improved.
b. In regard to earlier work, adding $LiBH_4$ together with $ZrFe_2$ is a suitable catalyst that took care of pure phase formation and improved the hydrogen sorption properties of 1:2 $Mg(NH_2)_2$-LiH. Using the additive LiBH4 to the above leads to the formation of the *in situ* catalyst $Li_4(BH_4)(NH_2)_3$. The synergistic effect of this quaternary catalyst and $ZrFe_2$ on the basic system 1:2 $Mg(NH_2)_2$-LiH can be analyzed and observed. This will lead to the improvement in hydrogen sorption characteristics.
c. The catalytic effect of carbon nanostructures in 1:2 $Mg(NH_2)_2$-LiH will be interesting to explore. If composites of CNTs and graphene can be

made and employed, it effectively enhances hydrogen sorption properties. The catalytic effect has to be expected through ballistic transport of hydrogen through nanotubes, which improve the hydrogen sorption of 1:2 $Mg(NH_2)_2$-LiH.

d. Further, if it can be attempted to replace $LiBH_4$ with a more abundant $NaBH_4$ system to see whether $NaBH_4$ on its own or in conjunction with $LiBH_4$ works as a better additive/catalyst for the basic 1:2 $Mg(NH_2)_2$-LiH storage system, it has been found that, unlike $LiBH_4$, $NaBH_4$ does not lead to the formation of in situ quaternary catalysts, but acts as a catalyst for enhancing hydrogen sorption. In the earlier research, composite 1:2 $Mg(NH_2)_2$-LiH, catalyzed by pulverizer cum catalyst (FeTi) and $Li_4(BH_4)(NH_2)_3+NaBH_4$, reversibly store H_2 (4.9 wt.%) at moderate temperature (170°C).

10.2.4 Ammonia Borane for Hydrogen Storage

It is now known that ammonia borane is a high gravimetric density hydrogen storage material [66,67]. The ammonia borane (NH_3BH_3) compound contains 19.6 wt.% of H_2, which in part can be brought to about 130°C by a multi-stage decomposition process [68,69]. A universal yield of 14% by weight of hydrogen is initiated in decomposition stages. The product of thermolysis is borazine. However, the experiments on dehydrogenation of ammonia borane (NH_3BH_3) confirm the traces of diborane and borazine on thermolysis of it in the hydrogen gas. The produced borazine and diborane are toxic products that make this material less promising. The dehydrogenation of ammonia borane molecules is similar to that of $NaBH_4$. The ammonia triborane ($NH_3B_3H_7$), with a hydrogen storage capacity of 17.8 wt.% [69], is an attractive material for onboard hydrogen storage. It releases eight moles of hydrogen during hydrolysis as per equation 10.12. The solubility of the ammonia triborane ($NH_3B_3H_7$) is 33% by weight in water. This is higher than the solubility of NH_3BH_3, which is 26% by weight. Noble metals can be used as catalysts for rapid hydrogen startup; rhodium complexes supported on alumina seem like a quality catalyst. According to the materials ($NH_3B_3H_7+H_2O+$Rh/alumina), 6.1 wt.% H_2 was emitted at ambient temperature (Figure 10.8).

$$NH_3B_3H_7+6H_2O \rightarrow NH_4{}^{+}+2H^{+}+3BO^{2-}+8H_2 \tag{10.12}$$

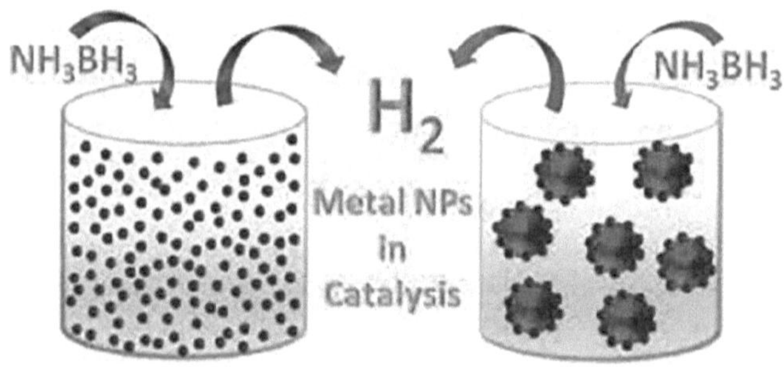

FIGURE 10.8 Ammonia borane as hydrogen storage materials [66].

More recently, the Co_3O_4 sheet has been found to have an improved catalytic effect on enhancing the hydrogen release rate during the hydrolysis of NH_3BH_3 [70]. Nonetheless, the same reservations apply to ammonia boranes to metal or metal hydride hydrolysis for hydrogen storage. The dehydrogenation and few properties of $NH_3B_3H_7$ are very similar to the $NaBH_4$ system. The millennium cell was introduced with the advantage of increased storage capacity, but with the disadvantage of possible contamination of the fuel cell with ammonia. For practical applications, new methods leading to the generation of borates from aqueous solutions are also needed. This technique requires very high energy. It seems almost impossible to reduce boron compounds to ammonia boron in an aqueous solution, requiring further isolation of boron oxide. At low temperatures, the infusion of ammonia borane into the mesoporous silica material SBA-15 can greatly improve hydrogen release kinetics (but not the thermodynamics) [66–70]. The infusion of porous carbon materials has been found promising, which can modify the re-/dehydrogenation kinetics of the material. But the disadvantage with it is that it significantly reduces the storage capacity of the composite material.

10.3 CONCLUSIONS

In the search for efficient materials for vehicular transportation, light complex metal hydrides have been described as promising materials for storing large amounts of hydrogen. Therefore, intensive research activities have been carried out over the decades. Metal borohydrides are increasingly in demand due to their high hydrogen storage capacity of 18.4 wt.%. Furthermore, the structural flexibility and different constituent compositions of these materials offer a very rich chemical composition. Different types of destabilizing approaches have been discussed in the present chapter to make viable hydrogen storage materials. The present chapter potentially highlights the lightweight complex metal hydrides for use in different applications. Based on the extensive knowledge of the new compound, it is expected to develop new metal borohydrides with the desired chemical composition, atomic coordination, and attractive properties.

ACKNOWLEDGMENTS

The authors would like to express their gratitude to CSIR, India, and BHU for giving a fellowship and a platform for their doctoral research.

NOTE

1 This book chapter is dedicated to our beloved supervisor and distinguished researcher in the area of hydrogen energy, Prof. O.N. Srivastava, who departed prematurely from this world on 24 April 2021.

REFERENCES

1. Ley, MB, Jepsen, LH, Lee, Y-S, Cho, YW, Bellosta von Colbe, JM, Dornheim, M, Rokni, M, Jensen, JO, Sloth, M, Filinchuk, Y, et al. Complex hydrides for hydrogen storage—New perspectives. *Mater. Today*, 17, 122–128, 2014.

2. Rude, LH, Nielsen, TK, Ravnsbæk, DB, Bösenberg, U, Ley, MB, Richter, B, Arnbjerg, LM, Dornheim, M, Filinchuk, Y, Besenbacher, F. Tailoring properties of borohydrides for hydrogen storage: A review. *Phys. Status Solidi A*, 208, 1754–1773, 2011.
3. Srivastav, ON, Yadav TP, Shahi RR, Pandey SK, Shaz MA, Bhatnagar A, Hydrogen energy in India: Storage to application. *Proc. Indian Natl. Sci. Acad.*, 81, 915–937, 2015.
4. Hudson MSL, Dubey PK, Pukazhselvan D, Pandey SK, Singh RK, Raghubanshi H, Hydrogen energy in changing environmental scenario: Indian context. *Int. J. Hydrog. Energy*, 34, 7358–67, 2009.
5. Bhatnagar A, Pandey SK, Vishwakarma AK, Singh S, Shukla V, Soni PK, Srivastava ON, Fe3O4@graphene as a superior catalyst for hydrogen de/absorption from/in MgH2/Mg. *J. Mat. Chem.*, 4, 14761–7, 2016.
6. Paskevicius, M, Jepsen, LH, Schouwink, P; Cerný, R; Ravnsbæk, DB; Filinchuk, Y; Dornheim, M; Besenbacher, F; Jensen, TR Metal borohydrides and derivatives – Synthesis, structure and properties. *Chem. Soc. Rev.*, 46, 1565–1634, 2017.
7. Rogelj, J., Luderer, G., Pietzcker, R. C., Kriegler, E., Schaeffer, M., Krey, V., & Riahi, K. (2015). Energy system transformations for limiting end-of-century warming to below 1.5 C. *Nature Climate Change*, *5*(6), 519–527.
8. Sanna, M., Ng, S., Vaghasiya, J. V., & Pumera, M. (2022). Fluorinated MAX Phases for Photoelectrochemical Hydrogen Evolution. *ACS Sustainable Chemistry & Engineering*, *10*(8), 2793–2801.
9. Bréon FM, Broquet G, Puygrenier V, Chevallier F, Xueref-Remy I, Ramonet M, Dieudonné E, Lopez M, Schmidt M, Perrussel O, Ciais P. An attempt at estimating Paris area CO_2 emissions from atmospheric concentration measurements. *Atm. Chem. Phys.*, 15, 1707–24, 2015.
10. Taub D. Effects of rising atmospheric concentrations of carbon dioxide on plants. *Nat. Educ. Knowl.*, 1, 2010.
11. Bogdanovic, B, Schwickardi, M. Ti-doped alkali metal aluminium hydrides as potential novel reversible hydrogen storage materials. *J. Alloys Compd.* 253–254, 1–9, 1997.
12. Züttel, A, Wenger, P, Rentsch, S, Sudan, P, Mauron, P, Emmenegger, C. LiBH4 a new hydrogen storage material. *J. Power Sources*, 118, 1–7, 2003.
13. Chen, P, Xiong, Z, Luo, J, Lin, J, Tan, KL. Interaction of hydrogen with metal nitrides and imides. *Nature*, 420, 302–304, 2002.
14. Møller KT, Sheppard D, Ravnsbæk DB, Buckley CE, Akiba E, Li HW, Jensen TR. Complex metal hydrides for hydrogen, thermal and electrochemical energy storage. *Energies*, 10, 1645, 2017.
15. Wang, J, Li, H-W, Chen, P. Amides and borohydrides for high-capacity solid-state hydrogen storage – Materials design and kinetic improvements. *MRS Bull.*, 38, 480–487, 2013.
16. Nakamori, Y, Kitahara, G, Orimo, S. Synthesis and dehydrating studies of Mg–N–H systems. *J. Power Sources*, 138, 309–312, 2004.
17. Smallbone A, Jia B, Atkins P, Roskilly AP. The impact of disruptive powertrain technologies on energy consumption and carbon dioxide emissions from heavy-duty vehicles. *Energy Convers. Manag.*, 6, 100030, 2020.
18. Wu Z, Wee V, Ma X, Zhao D. Adsorbed natural gas storage for onboard applications. *Adv. Sustain. Syst.*, 2000200, 2021.
19. Ouyang L, Chen K, Jiang J, Yang XS, Zhu M. Hydrogen storage in light-metal based systems: A review. *J. Alloys Comp.*, 829, 154597, 2020.
20. Møller KT, Sheppard D, Ravnsbæk DB, Buckley CE, Akiba E, Li HW, Jensen TR. Complex metal hydrides for hydrogen, thermal and electrochemical energy storage. *Energies*. 2017 Oct 18;10(10):1645.

21. Rzeszotarska M, Czujko T, Polański M. Mg2 (Fe, Cr, Ni) HX complex hydride synthesis from austenitic stainless steel and magnesium hydride. *Int. J. Hydrogen Energy*, 45, 19440–54, 2020.
22. Lai Q, Pratthana C, Yang Y, Rawal A, Aguey-Zinsou KF. Nanoconfinement of complex borohydrides for hydrogen storage. *ACS Appl. Nano Mater.*, 4, 973–8, 2021.
23. Ali NA, Ismail M. Modification of NaAlH4 properties using catalysts for solid-state hydrogen storage: A review. *Int. J. Hydrogen Energy*, 2020.
24. Leng, HY, Ichikawa, T, Hino, S, Hanada, N, Isobe, S, Fujii, H. New metal–N–H system composed of Mg(NH2)2 and LiH for hydrogen storage. *J. Phys. Chem.*, 108, 8763–8765, 2004.
25. Luo, W. (LiNH2–MgH2): A viable hydrogen storage system. *J. Alloys Comp.*, 381, 284–287, 2004.
26. Xiong, Z, Hu, J, Wu, G, Chen, P, Luo, W, Gross, K, Wang, J. Thermodynamic and kinetic investigations of the hydrogen storage in the Li–Mg–N–H system. *J. Alloys Comp.*, 398, 235–239, 2005.
27. Torre, F, Valentoni, A, Milanese, C, Pistidda, C, Marini, A, Dornheim, M, Enzo, S, Mulas, G, Garroni, S. Kinetic improvement on the CaH2-catalyzed Mg(NH2)2 + 2LiH system. *J. Alloys Comp.*, 645, S284–S287, 2015.
28. Shaw, LL, Ren, R, Markmaitree, T, Osborn, W. Effects of mechanical activation on dehydrogenation of the lithium amide and lithium hydride system. *J. Alloys Comp.*, 448, 263–271, 2008.
29. Liu, Y, Zhong, K, Luo, K, Gao, M, Pan, H, Wang, Q. Size-dependent kinetic enhancement in hydrogen absorption and desorption of the Li–Mg–N–H system. *J. Am. Chem. Soc.*, 131, 1862–1870, 2009.
30. Orimo SI, Nakamori Y, Eliseo JR, Züttel A, Jensen CM. Complex hydrides for hydrogen storage. *Chem. Rev.*, 107, 4111–32, 2007.
31. Züttel A, Rentsch S, Fischer P, Wenger PM, Sudan PH, Mauron P, Emmenegger C. Hydrogen storage properties of LiBH4. *J. Alloys Comp.*, 356, 515–20, 2003.
32. Gross AF, Vajo JJ, Van Atta SL, Olson GL. Enhanced hydrogen storage kinetics of LiBH4 in nanoporous carbon scaffolds. *J. Phys. Chem.*, 112, 5651–7, 2008.
33. Jin SA, Lee YS, Shim JH, Cho YW. Reversible hydrogen storage in LiBH4–MH2 (M=Ce, Ca) composites. *J. Phys. Chem.*, 112, 9520–4, 2008.
34. Matsunaga T, Buchter F, Miwa K, Towata S, Orimo S, Züttel A. Magnesium borohydride: A new hydrogen storage material. *Renew. Energy*, 33, 193–6, 2008.
35. Bououdina M. Nanomaterials for hydrogen storage: Renewable and clean energy. *Int. J. Nanoelectron. Mater.*, 3, 155–67, 2010.
36. Escobar D. *Investigation of ZrNi, ZrMn 2 and Zn (BH 4) 2 Metal/Complex Hydrides for Hydrogen Storage*, 2007.
37. PI JC, Cui J, Gao Y, Kniajansky S, Lemmon J, Raber T, Rijssenbeek J, Rubinsztajn G, Soloveichik G. *Lightweight Intermetallics for Hydrogen Storage*, 2004.
38. Rönnebro E, Majzoub EH. Calcium borohydride for hydrogen storage: Catalysis and reversibility. *J. Phys. Chem.*, 2007, 111(42), 12045–7.
39. Milanese C, Garroni S, Gennari F, Marini A, Klassen T, Dornheim M, Pistidda C. Solid state hydrogen storage in alanates and alanate-based compounds: A review. *Metals*, 2018, 8(8), 567.
40. Fichtner M, Engel J, Fuhr O, Kircher O, Rubner O. Nanocrystalline aluminium hydrides for hydrogen storage. *Mater. Sci. Eng.*, 2004, 108(1–2), 42–7.
41. von Colbe JM, Lozano G, Metz O, Bücherl T, Bormann R, Klassen T, Dornheim M. Design, sorption behaviour and energy management in a sodium alanate-based lightweight hydrogen storage tank. *Int. J. Hydrogen Energy*, 2015, 40(7), 2984–8.

42. Yin LC, Wang P, Kang XD, Sun CH, Cheng HM. Functional anion concept: effect of fluorine anion on hydrogen storage of sodium alanate. *Phys. Chem. Chem. Phys.*, 2007, 9(12), 1499–502.
43. Dehouche Z, Lafi L, Grimard N, Goyette J, Chahine R. The catalytic effect of single-wall carbon nanotubes on the hydrogen sorption properties of sodium alanates. *Nanotechnology*, 2005, 16(4), 402.
44. Jiang R, Xiao X, Zheng J, Chen M, Chen L. Remarkable hydrogen absorption/desorption behaviors and mechanism of sodium alanates in-situ doped with Ti-based 2D MXene. *Mater. Chem. Phys.*, 2020, 242, 122529.
45. Beatrice CA, Moreira BR, de Oliveira AD, Passador FR, de Almeida Neto GR, Leiva DR, Pessan LA. Development of polymer nanocomposites with sodium alanate for hydrogen storage. *Int. J. Hydrogen Energy*, 2020, 45(8), 5337–46.
46. Milanese C, Garroni S, Gennari F, Marini A, Klassen T, Dornheim M, Pistidda C. Solid state hydrogen storage in alanates and alanate-based compounds: A review. *Metals*, 2018, 8(8), 567.
47. Srinivasan SS, Brinks HW, Hauback BC, Sun D, Jensen CM, Long term cycling behavior of titanium doped NaAlH4 prepared through solvent mediated milling of NaH and Al with titanium dopant precursors. *J. Alloys Comp.*, 377, 283–289, 2004.
48. Bhatnagar A, Pandey SK, Dixit V, Shukla V, Shahi RR, Shaz MA, Srivastava ON, Catalytic effect of carbon nanostructures on the hydrogen storage properties of MgH_2–$NaAlH_4$ composite. *Int. J. Hydrogen Energy*, 39, 14240–6, 2014.
49. Bogdanović B, Brand RA, Marjanović A, Schwickardi M, Tölle J. Metal-doped sodium aluminium hydrides as potential new hydrogen storage materials. *J. Alloys Comp.*, 302, 36–58, 2000.
50. Jensen CM, Sun D, Srinivasan S, Wang P, Murphy K, Wang Z, Kuba M, Sulic M, Richards A, Linzi J, Niemczura W. *Catalytically Enhanced Hydrogen Storage System, DOE Progress Report*, 2004.
51. Chen P, Xiong Z, Luo, J, Tan J, Lin KL. Interaction between lithium amide and lithium hydride. *J. Phys. Chem.*, 107, 10967–10970, 2003.
52. Chen P, Xiong Z, Luo, J, Tan J, Lin KL. Lithium borohydride hydrogen storage material decorated by oxide and preparation method thereof. *Nature*, 420, 302–309, 2002.
53. Garroni S, Santoru A, Cao H, Dornheim M, Klassen T, Milanese C, Gennari F, Pistidda C. Recent progress and new perspectives on metal amide and imide systems for solid-state hydrogen storage. *Energies*, 2018, 5, 1027.
54. Shukla V, Bhatnagar A, Pandey SK, Shahi RR, Yadav TP, Srivastava ON. On the synthesis, characterization and hydrogen storage behavior of ZrFe2 catalyzed Li–Mg–N–H hydrogen storage material. *Int. J. Hydrogen Energy*, 40, 12294–302, 2015.
55. Shukla V, Bhatnagar A, Singh S, Soni PK, Verma SK, Yadav TP, Shaz MA, Srivastava ON. A dual borohydride (Li and Na borohydride) catalyst/additive together with intermetallic FeTi for the optimization of the hydrogen sorption characteristics of Mg (NH 2) 2/2LiH. *Dalton Transac.*, 48, 11391–403, 2019.
56. Shukla V, Bhatnagar A, Soni PK, Vishwakarma AK, Shaz MA, Yadav TP, Srivastava ON. Enhanced hydrogen sorption in a Li–Mg–N–H system by the synergistic role of Li 4 (NH 2) 3 BH 4 and ZrFe 2. *Phys. Chem. Chem. Phys.*, 19, 9444–56, 2017.
57. Shahi RR, Yadav TP, Shaz, MA, Srivastava ON. Studies on dehydrogenation characteristic of Mg (NH2) 2/LiH mixture admixed with vanadium and vanadium based catalysts (V, V2O5 and VCl3). *Int. J. Hydrogen Energy*, 35, 238–246, 2010.
58. Hino S, Nakagawa T, Fujii H. Mechanism of hydrogenation reaction in the Li–Mg–N–H system. *J. Phys. Chem. B.*, 109, 10744–8, 2005.

59. Wu G, Xiong, Z, Liu, T, Liu, Y, Hu, J, Chen, O, Feng, Y, Wee, ATS. Synthesis and characterisation of a new ternary imide – Li2Ca(NH)2. *Inorganic Chemistry*, 46, 517–521, 2007.
60. Vajo JJ, Skeith SL, Mertens F. Reversible storage of hydrogen in destabilized LiBH4. *J. Phys. Chem. B*, 109, 3719–3722, 2005.
61. Filinchuk YE, Yvon K, Meisner GP, Pinkerton FE, Balogh MP. On the composition and crystal structure of the new quaternary hydride phase Li4BN3H10. *Inorg. Chem.*, 45, 1433–1435, 2006.
62. Luo W. (LiNH2-MgH2): A viable hydrogen storage system. *J. Alloys Comp.*, 381, 284–287, 2004.
63. Xiong Z, Wu GU, Hu J, Chen PI. Ternary imides for hydrogen storage. *Adv. Mater.*, 16, 1522–5, 2004.
64. Sudik A, Yang J, Halliday D, Wolverton C, Kinetic improvement in the Mg(NH2)2–LiH storage system by product seeding. *J. Phys. Chem.*, 111, 6568–73, 2007.
65. Yang J, Sudik A, Siegel DJ, Halliday D, Drews A, Carter IIIRO, Wolverton C, Lewis GJ, Sachtler JWA, Low JJ, Faheem SA, Lesch DA, Ozolins V, Hydrogen storage properties of 2LiNH2+ LiBH4+ MgH2. *J. Alloys Comp.*, 446, 345–349, 2007.
66. Stephens FH, Pons V, Baker RT, Ammonia–borane: The hydrogen source par excellence?. *Dalton Transac.*, 25, 2613–26, 2007.
67. Peng B, Chen J, Ammonia borane as an efficient and lightweight hydrogen storage medium. *Energy Environ. Sci.*, 1, 479–83, 2008.
68. Akbayrak S, Özkar S, Ammonia borane as hydrogen storage materials. *Int. J. Hydrogen Energy*, 43, 18592–606, 2018.
69. Yamada Y, Yano K, Xu QA, Fukuzumi S, Cu/Co3O4 nanoparticles as catalysts for hydrogen evolution from ammonia borane by hydrolysis. *J. Phys. Chem. C*, 114, 16456–16462, 2010.
70. Yoon CW, Carroll PJ, Sneddon LG, Ammonia triborane: A new synthesis, structural determinations, and hydrolytic hydrogen-release properties. *J. Am. Chem. Soc.*, 131, 855–64, 2009.

11 Assessing the Feasibility of Floating Photovoltaic Plant at Mukutmanipur in India

D. Misra
Techno India Group

CONTENTS

11.1 INTRODUCTION

In recent times, the energy demand is continuously increasing due to the rapid growth of population and its consequence is rapid diminishing of fossil fuels, which leads to the shift of our focus to renewable energy resources. To ensure new renewable energy for sustainable future, many developing countries are shifting towards floating solar photovoltaic (FPV) system [1]. A FPV system comprises solar panels and floating structural arrangement, which are installed on water surface. Solar panels are affixed to the buoyant floating structure that keeps them on top of the water surface. Generally, stagnant water bodies such as lakes, dam water reservoirs, water treatment plants, municipality water storage ponds, hydropower reservoirs where the water surface is still or calmer are the best options for FPV system installation [2]. However, river or sea surface can also be considered by installing specially designed floating structure. Although the FPV technology was started from 2008, it is continuously expanding across the world as it is an appropriate solution of renewable energy generation in situations in which roof and/or ground space is limited.

India is a country with enormous solar energy potential, about 4–7 kWh/m^2 solar radiation falls in a day. Thus, solar energy could meet the huge energy demand of the country. As the arable land in India is continuously depleted due to the fast-growing

DOI: 10.1201/9781003272717-11

population and waste land is also scarce here, the FPV technology is the best option to utilize the unutilized water bodies, which not only preserves the agricultural lands, but also conserves water by controlling the water evaporation rate [3]. Indian states such as Andhra Pradesh, Tamil Nadu, Karnataka, Kerala, Madhya Pradesh, Maharashtra, and West Bengal possess several numbers of natural and man-made water bodies, which can be effectively utilized for harvesting solar energy potential. India, owing to its inherent geographical benefits, can develop large-scale FPV projects. Therefore, a vast land area will not be required for implementing the projects as required for a land-based PV system. Indian Government recently planned to install large-scale FPV plants by 2020–2021.

Ravichandran et al. [4] proposed and presented a comprehensive numerical model for an FPV system in Mettur reservoir (Tamil Nadu), India. Their proposal was based on different inclinations of PV panels as well as tracking mechanisms. It was found that the FPV plant could save 184589 m^3 of water evaporation in a year. Tina et al. [5] developed mathematical models that could estimate the performances of bifacial and monofacial PV panels fitted on floated planes on water. The models were validated acquiring data from experimental FPV set-ups developed in the Enel Innovation Lab, Catania, Italy. Celemons et al. [6] presented a life cycle and cost-benefit analysis of FPV plants considering the 150 MW plant in Thailand. They also analysed FPV plant performances taking 21 reservoirs with change in the installation coverage areas. They also compared levelized cost of energy for ground-based and water-based PV power generation.

Aid et al. [7] expressed the importance of FPV plants for areas with acute water scarcity in Central and South Asia. They carried out a review study showing technical and economic feasibility of the FPV system. Choi and Lee [8] reported that wind and wave load should be carefully estimated to provide stability to an ocean FPV system. They efficiently designed suitable floating structures considering appropriate load and strength estimation for the completion of a 20 kW ocean FPV plant. Sacramento [9] made a comparative study between ground-mounted PV and FPV systems in Brazil. They found that the average conversion efficiency is increased by 12.5% by using the FPV system compared to the ground-mounted PV system. Ferrer-Gisbert et al. [10] presented a new technique to develop an FPV system to prevent water evaporation loss in reservoirs, which store water for agriculture irrigation purpose at Alicante in Spain. Here, medium-density polyethylene (MDPE) floating modules/structures, which had a boat-like shape, were chosen to support FPV. The modules were joined together with a metallic pin-anchorage and could float any water level. They also found that the system was economically viable although the system cost was 30% higher than the regular PV system cost. Choi [11,12] demonstrated a comparative study on the power generation and environmental impact between the floating and ground-mounted PV system. He studied different FPV power generation units of 2.4, 100, and 500 kW in Hapcheon Dam water reservoir (Korea Water Resources Corporation). He also mentioned that the wind velocity, water flow, and solar radiation were the influential parameters on the performance of the FPV system. Choi et al. [13] stated the technique to develop a 100 kW FPV power plant at Hapcheon Dam reservoir in Korea. In their work, the design strategy for the system equipment such as floating structure, mooring device, water cable, and control unit were suitably presented to resist natural calamities. Durkovic and Durisic [14] presented a conceptual study of the FPV plant on Skadar Lake, situated in Albania and Montenegro border. They used various environmental data and numerous mathematical equations for designing FPV system and finally concluded that the system could be economically viable for generating power. Kim et al. [15] designed and constructed a large-scale FPV plant by using a high-capacity floating structure made of steel, aluminium, and fibre-reinforced polymer (FRP). They stated that owing to FRP fabrication, the system was light in weight and economically viable. Mittal et al. [16] carried out a comparative study between a 1 MW FPV plant and a 1 MW regular PV plant in a specific location in India. They found that the annual generation capacity of FPV plant was 2.48% higher than that of the regular PV plant. It was reported that the FPV system could reduce average 14.56% temperature of the module compared to regular PV system. Yearly performance ratio and capacity utilization factor were also slightly higher than those of the regular PV system.

In this chapter, a brief overview of the FPV technology and its effectiveness as a new kind of renewable energy has been presented. Finally, a study related to the feasibility of installing a FPV plant at Mukutmanipur reservoir has been illustrated. The reservoir was especially developed to conserve water for irrigating agricultural land as water becomes scarce in this area during summer. This site has a semi-arid climatic condition. Again, Mukutmanipur is pervaded with vibrant colour of Palash (*Butea monosperma*) and Sonajhuri (*Acacia auriculiformis*) trees, which attract tourists during spring. Already, this place is opened for tourists to promote tourism as well as a number of schemes have been taken up for beautifying the dam area. Thus, this area could be further improved by installing an FPV plant on the reservoir, which can serve multiple functions such as power generation, beautification with musical fountain show, and water saving by managing evaporation loss. Again, the FPV plant with fountain could improve water aeration without hampering the aquaculture habitat of a reservoir.

11.2 PROPOSED SITE DETAIL

In West Bengal, most of the large-scale solar projects have been established in the south and south-west part of the state owing to the high solar radiation on these areas. Recently, a 10 MW solar photovoltaic plant has been completed at Mejia, Bankura. Three solar projects, each having 10 MW capacity, have also been conceived in Purulia and Bankura [17,18]. Mukutmanipur is also located at the south-west part of Bankura in West Bengal and receives ample solar radiation.

11.2.1 Site Selection Criteria

In the development of an FPV plant, the selection of suitable site is very important as stability and power generation of the plant depend on weather conditions of the site. Solar panels support system on floating structures is influenced by a variety of environmental factors such as geographical conditions and regional wind speed. Thus, the installation cost could vary depending on the condition of the site. Following factors should be taken into consideration for the installation of an FPV plant:

i. Things such as solar radiation level, sunshine hours, fog and mist, and shade by tree, which directly affect the power generation.

ii. Depth of water in the reservoir throughout the year, inflow of waste floating particles, accessibility of water for other purpose, etc.
iii. Distance from power station or substation for supplying power, which may affect the transmission and distribution costs.
iv. Legal restriction and acts towards water security, environmental preservation, river acts, special protection of fauna and flora, civil complaints, etc.

11.2.2 Proposed Site Background

Mukutmanipur is located in the south-west part of West Bengal (22.95° N, 86.75° E) and situated at an elevation of 94 m above sea level. It is in Bankura district of West Bengal, India. The reservoir is located at the confluence of two rivers – Kangsabati and Kumari, surrounded by hills, forests, deer park, islands with variety of migratory birds, etc. Mukutmanipur dam is the India's second longest man-made earthen dam, which is 11 km long and 38 m high. There is a water reservoir around it, which is 124.32 km^2 in area. The dam was primarily built to provide irrigational facilities to the arable land of 8,000 km^2, covering Purulia, Bankura, Paschim Medinipur, and some parts of West Hooghly (Table 11.1).

Climatically, Mukutmanipur lies in the tropical zone and is characterized by hot and dry weather. The temperature of the site is almost moderate during the winter and significantly rises from the beginning of March. The peak temperature of the site can be more than 40°C. Generally, monsoon starts from the first week of June and the temperature drops appreciably. In winter, the mean temperature rarely falls below 7°C. The relative humidity is high in between June and October. Sky is mostly clouded in June–August due to south-west monsoon, and it is almost clear in the rest of the year. Annual mean wind speed ranges in between 2.6 and 7.1 km/h. The wind speed is higher in April–June compared to other months. Storms and depressions from the Bay of Bengal in May and in the post-monsoon period often cause widespread heavy rain associated with strong winds (Figures 11.1 and 11.2).

An assessment of the solar radiation is the first criterion for initiating a plant related to any solar project. In the proposed site, annual average global solar radiation on a horizontal surface is 5.35 kWh/m^2/day. Solar radiation is minimum in December, and February–June are the high solar radiation months and the remaining months get almost average solar radiation. Figure 11.3 shows the solar radiation data along

TABLE 11.1
Details of the Water Reservoir [19]

Parameter	Value
Area of reservoir	124.32 km^2
Depth of lowest water level	15 m
Optimum water level	134 m
Catchment area	3625 km^2
Maximum area for irrigation	274940 ha

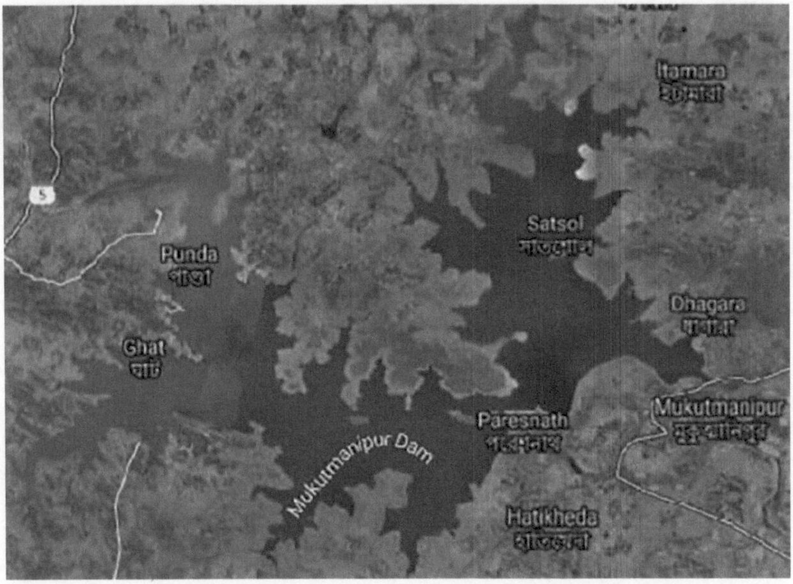

FIGURE 11.1 Satellite image of the water reservoir at Mukutmanipur [20].

FIGURE 11.2 Water reservoir at Mukutmanipur [21].

with wind speed. In the proposed site, the wind velocity is moderate throughout the year and solar radiation is also within acceptable range to run the plant (Table 11.2).

Some other suitable criteria, which show satisfactory parameters for selecting the proposed site for the FPV plant, are as follows:

a. The site is close to a national highway (NH314).
b. The proposed site is a tourist spot, and power transmission systems are already available. Also, this site is near the Khatra Substation.

TABLE 11.2
Climatic Data of Mukutmanipur [21,22]

Month	Max. Temp. (°C)	Min. Temp. (°C)	RH (%)	Precipitation (mm)
January	26	12	60	12
February	29	16	48	15
March	34	20	45	22
April	38	23	38	30
May	40	26	53	58
June	37	27	70	201
July	33	26	83	293
August	31	25	83	277
September	30	29	82	222
October	31	21	74	93
November	31	18	62	12
December	27	15	59	14

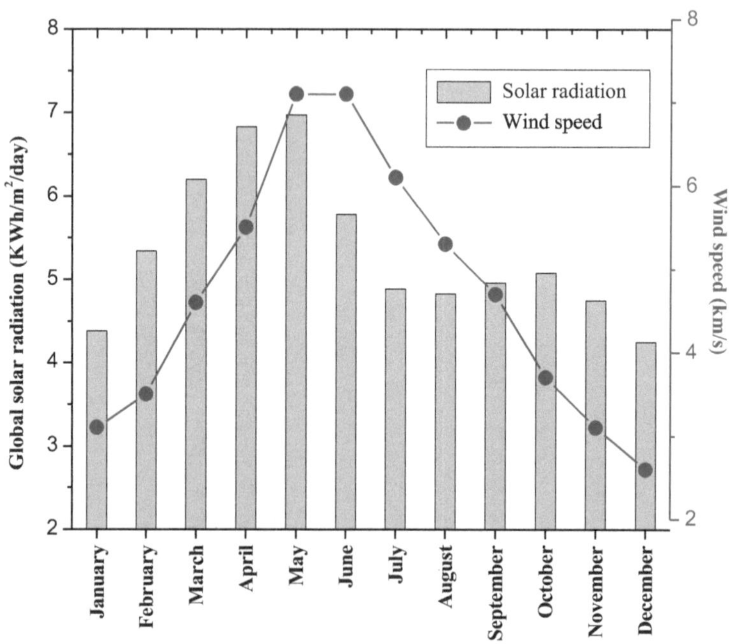

FIGURE 11.3 Solar radiation and wind speed in different months [22,23].

c. This location is not near any sensitive sites such as reserved forest and military cantonments.
d. The place is a seismic hazard-free zone.
e. The place is an open reservoir, and no tall tree is seen around it; generally, panel soiling by bird droppings would not occur.

In West Bengal, the first off-grid solar power plant was installed in Bankura in 1992. Again, in 2014, India's first FPV plant was installed at Rajarhat, Kolkata, in West Bengal [24]. It was 10 kW in capacity. However, the state is lagging behind the other Western and Southern states of the country for developing solar PV plant although the major portions of the state receive plenty of solar radiation [25]. In view of the improving momentum towards the solar PV generation in the national perspective, the government has started lots of initiatives in recent time and has set up a separate Solar Power Generation Department in West Bengal. Here, in West Bengal, the availability of barren or uncultivated land is very minimum and thus land is scarce. As the cost of the land is much higher here than that of the other solar power enriched states, land-based solar power generation is a very challenging task. From this point of view, Mukutmanipur is an ideal place where floating solar power plant can be built considering all favourable factors for the installation of the plant.

11.3 FEASIBILITY OF FPV PLANT AT THE PROPOSED SITE

After studying the literature, it is found that a 1 kW FPV system requires a maximum of 40 m^2 area of water surface for PV installation. Also, it is found that 10%–25% area of the water body is required for the installation of a plant. The present site could be used for installing a 310 MW floating PV plant considering only 10% area of the reservoir [3].

Energy generation calculation: Energy generated by the FPV plant depends on the daily average global solar radiation. The daily energy generation (E_{PV}) is calculated by the following equation.

$$E_{PV} = PR \times A_{array} \times \eta_p \times I_g \quad (11.1)$$

where PR is the performance ratio, A_{array} is the total PV panel area, η_p is the solar panel electrical efficiency, and I_g is the global solar radiation (kWh/m^2day) falling on the array. The PR of the present case is considered to be 0.78 [26,27].

Water-saving calculations: The Central Water Commission Basin Planning and Management Organization has developed a calculation for water saving for the regions located at the east of 87° E longitude, which include parts of West Bengal and the entire Northeast India. The yearly saving of water is calculated accordingly, and it is found that 1750 L/m^2/year of water could be saved due to the reduction in evaporation [28].

Calculation for reduction in CO_2 emissions: According to the Central Electricity Authority (India), in a coal-based thermal power plant, the CO_2 emission is 0.82 t/MWh [29]. Considering this, the reduction in CO_2 emissions in this 310 MW floating PV plant is estimated.

$$CO_2(\text{Emission}) = E_y \times 1000 \times 0.82 \quad (11.2)$$

where E_y is the total annual power generation in GWh.

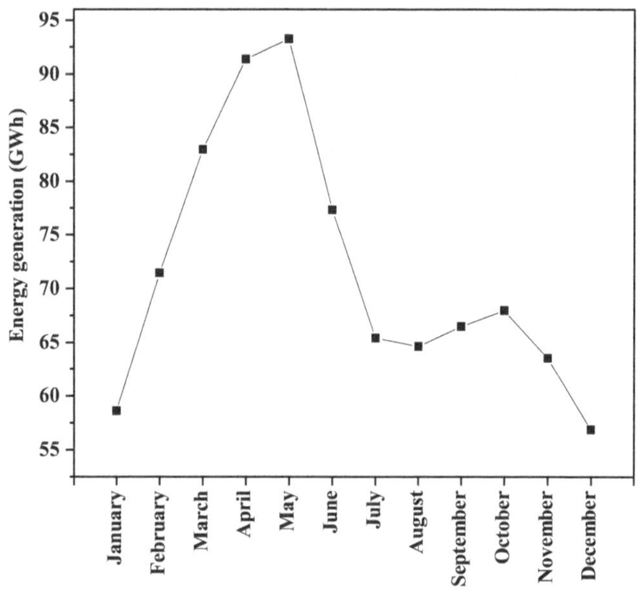

FIGURE 11.4 Monthly expected power generation by the plant.

Figure 11.4 shows the monthly power generation by the plant taking relevant solar radiation data for the site. It is observed that from March to June, the power generation is maximum compared to other months. The total annual power generation capacity of the plant would be 860 GWh.

11.4 ECONOMIC FEASIBILITY

The investment cost of FPV plants are more than that of ground-based PV plants, owing to some extra arrangement such as floats, moorings, and waterproof electrical components. The cost of floats depends on the materials used for the installation of the plant. From the last few years, the prices of the products used as material have been gradually decreasing. In India, in 2018, the total capital cost for an FPV plant installation was within the range of 1–1.2 USD per Wp depending upon the site, reservoir water depth, and total installed capacity [30]. Table 11.3 shows the breakup of different costs in Indian conditions.

The present FPV plant installation cost would be ₹ 2542 crores. It is found that a 310 MW FPV plant could generate 860 GWh/year energy valuing ₹ 559 crore/year considering the current average unit cost of electricity tariff ₹ 6.5/kWh [32]. It is considered that the yearly operation and maintenance cost for the plant is 2% of the capital cost. Thus, the simple payback period of the plant would be within 5 years.

To develop a 310 MW land-based solar photovoltaic plant, 1240 acres of vacant land is required. Land is limited and scarce in West Bengal. A few amount of infertile land is available here, and land price is also very high. However, a 10 MW solar power plant at Mejhia, Bankura, has been developed utilizing 178.9 acres of vested

TABLE 11.3
FPV System's Expected Cost [26,30,31]

S. No.	Item	Rating	Cost/unit (₹)	Total Cost (Crores)
1.	Solar modules	310×10^6 W	45/W	1395
2.	Floating structures	310×10^6 W	5/W	155
3.	Civil work	310×10^6 W	12/W	372
4.	Balance of system (BOS)	310×10^6 W	5/W	155
5.	Cables	310×10^6 W	4/W	124
6.	Power transmission set-up	310×10^6 W	3/W	93
7.	Miscellaneous	310×10^6 W	8/W	248

land [33]. The land price in the district of Bankura is about ₹ 6.34 lakh/acre [34]. Thus, the acquisition of 1240 acres of land for the plant will be a cost burden of ₹ 78.61 crores. With that huge amount of land price, the unit cost of electricity also increases. Thus, the present FPV plant could also save ₹ 78.61 crores.

11.5 CONCLUSIONS

The present study delineates that Mukutmanipur has enough potentiality to develop an FPV plant. A feasibility study of the FPV plant has been conducted for the site, which could be able to meet 345 GWh load demand per annum. Again, by saving the vacant land, the proposed plant is saving the huge amount of land cost, which could reduce the electric unit cost as well. However, in the long run, further profound knowledge should be gained about the reservoir water conservation and different cost-benefit analyses. Again, the plant has a positive impact on the environment. The proposed plant could save not only the emission of CO_2 of 705200 tonnes/year, but also 21756 million L of water by decreasing the water evaporation rate. Thus, if this proposed project of the FPV plant at Mukutmanipur gets enough legal and economic support as well as support from the government for implementation, it would be quite beneficial for the surrounding areas. Furthermore, with strong determination in the application of solar photovoltaic and a massive scale of exposure by the state government, it could ensure developing FPV plants in future.

REFERENCES

[1] A. Sahu, N. Yadav, K. Sudhakar, Floating photovoltaic power plant: A review, *Renewable and Sustainable Energy Reviews*, 66, 815–824, 2016.

[2] A. Kumar, I. Purohit, T. C. Kandpal, Assessment of floating solar photovoltaic (FSPV) potential in India, In *Proceedings of the 7th International Conference on Advances in Energy Research*, Springer, Singapore, pp. 973–982, 2021.

[3] D. Misra, Floating photovoltaic plant in India: Current status and future prospect, In *International Conference on Thermal Engineering and Management Advances*, Springer, Singapore, pp. 219–232, 2020.
[4] N. Ravichandran, N. Ravichandran, B. Panneerselvam, Performance analysis of a floating photovoltaic covering system in an Indian reservoir, *Clean Energy*, 5(2), 208–228, 2021.
[5] G. M. Tina, F. B. Scavo, L. Merlo, F. Bizzarri, Analysis of water environment on the performances of floating photovoltaic plants, *Renewable Energy*, 175, 281–295, 2021.
[6] S. K. C. Clemons, C. R. Salloum, K. G. Herdegen, R. M. Kamens, S. H. Gheewala, Life cycle assessment of a floating photovoltaic system and feasibility for application in Thailand, *Renewable Energy*, 168, 448–462, 2021.
[7] M. Abid, Z. Abid, J. Sagin, R. Murtaza, D. Sarbassov, M. Shabbir, Prospects of floating photovoltaic technology and its implementation in Central and South Asian Countries, *International Journal of Environmental Science and Technology*, 16(3), 1755–1762, 2019.
[8] Y. K. Choi, J. H. Lee, Structural safety assessment of ocean-floating photovoltaic structure model, *Israel Journal of Chemistry*, 55(10), 1081–1090, 2015.
[9] E. M. do Sacramento, P. C. Carvalho, J. C. de Araújo, D. B. Riffel, R. M. da Cruz Corrêa, J. S. P. Neto, Scenarios for use of floating photovoltaic plants in Brazilian reservoirs, *IET Renewable Power Generation*, 9(8), 1019–1024, 2015.
[10] C. Ferrer-Gisbert, J. J. Ferrán-Gozálvez, M. Redón-Santafé, P. Ferrer-Gisbert, F. J. Sánchez-Romero, J. B. Torregrosa-Soler, A new photovoltaic floating cover system for water reservoirs, *Renewable Energy*, 60, 63–70, 2013.
[11] Y. K. Choi, J. H. Lee, Structural safety assessment of ocean-floating photovoltaic structure model, *Israel Journal of Chemistry*, 55(10), 1081–1090, 2015.
[12] Y. K. Choi, A study on power generation analysis of floating PV system considering environmental impact, *International Journal of Software Engineering and Its Applications*, 8, 75–84, 2014.
[13] Y. K. Choi, N. H. Lee, A. K. Lee, K. J. Kim, A study on major design elements of tracking-type floating photovoltaic systems, *International Journal of Smart Grid and Clean Energy*, 3(1), 70–74, 2014.
[14] V. Durkovic, Z. Durisic, Analysis of the potential for use of floating PV power plant on the Skadar Lake for electricity supply of aluminium plant in Montenegr, *Energies*, 10(-10), 1505, 2017.
[15] S. H. Kim, S. J. Yoon, W. Choi, Design and construction of 1 MW class floating PV generation structural system using FRP members, *Energies* 10(8), 1142, 2017.
[16] D. Mittal, B. K. Saxena, K. V. S. Rao, Potential of floating photovoltaic system for energy generation and reduction of water evaporation at four different lakes in Rajasthan, *International Conference On Smart Technologies For Smart Nation (SmartTechCon)*, IEEE, pp. 238–243, 2017.
[17] Solar Projects, Department of Power Government of West Bengal. https://wbpower.gov.in/solar-projects/.
[18] West Bengal Renewable Energy Development Agency, http://www.wbreda.org/.
[19] Overview of Mukutmanipur. https://www.mukutmanipurtourism.com/overview-of-mukutmanipur/ (accessed March 10, 2020).
[20] SeattleiteimageofMukutmanipur. https://www.google.com/maps/place/Mukutmanipur+Dam. (accessed February 15, 2020).
[21] Mukutmanipur – Travel guide, https://en.wikivoyage.org/wiki/Mukutmanipur (accessed February 15, 2020).
[22] Average global solar radiation. http://www.synergyenviron.com/tools/solar-irradiance/Mukutmanipur (accessed February 20, 2020).

[23] Mukutmanipur climate and weather. https://www.holidayiq.com/destinations/mukutmanipur/weather.html (accessed March 1, 2020).

[24] Floating solar power plant in West Bengal, India. https://www.vikramsolar.com/case-studies/floating-solar-power-plant-in-west-bengal-india/ (accessed March 1, 2020).

[25] West Bengal lag in solar power promotion. https://timesofindia.indiatimes.com/city/kolkata/West-Bengal-lags-in-solar-power-promotion/articleshow/48138012.cms (accessed March 20, 2020).

[26] D. Misra, Design of a stand-alone rooftop PV system for electrification of an academic building, *International Journal of Engineering and Advanced Technology* 9(2), 3955–3964, 2019.

[27] A. Ganguly, D. Misra, S. Ghosh, Modeling and analysis of solar photovoltaic-electrolyzer-fuel cell hybrid power system integrated with a floriculture greenhouse, *Energy and Buildings*, 42(11), 2036–2043, 2010.

[28] CWC, *Evaporation Control in Reservoirs*. Basin Planning & Management Organisation, Central Water Commission, New Delhi, 2011 (accessed March 1, 2020).

[29] *Central Electricity Regulatory Commission, Ministry of Power, Government of India.* https://cercind.gov.in/advice_gov.html (accessed February 20, 2020).

[30] World Bank Group, ESMAP, and SERIS, *Where Sun Meets Water: Floating Solar Market Report—Executive Summary*. World Bank, Washington, DC, 2018. http://documents.worldbank.org/curated/en/579941540407455831/pdf/Floating-Solar-Market-Report-Executive-Summary.pdf (accessed March 20, 2020).

[31] A. Goswami, P. Sadhu, U. Goswami, P. K. Sadhu, Floating solar power plant for sustainable development: A techno economic analysis, *Environmental Progress & Sustainable Energy*, 38(6), 13268, 2019.

[32] West Bengal State Electricity Distribution Company Limited, https://www.wbsedcl.in/irj/go/km/docs/internet/new_website/Home.html (accessed March 20, 2020).

[33] Performance Audit Implementation of Renewable Energy Programme in West Bengal. https://saiindia.gov.in/uploads/download_audit_report/2018/Chapter_2_Performance_Audit_of_Report_No_4_of_2018_Economic_Sector_Government_of_West_Bengal.pdf.

[34] Directorate of Registration and Stamp Revenue, West Bengal. https://wbregistration.gov.in/(S(pbiiro25xztjkt3iswtw3dkg))/index.aspx (accessed February 20, 2020).

12 Floating Photovoltaic Systems
An Emerging PV Technology

Manish Kumar, Roopmati Meena, and Rajesh Gupta
Indian Institute of Technology Bombay

CONTENTS

12.1 INTRODUCTION

Increased technological advances maximize PV deployment and electricity generation, but the primary issue that must be addressed is the use of a vast area for electricity generation when multi-megawatt deployments are made. This has opened the door for the creation of a new technology known as floating photovoltaic (FPV), which focuses on water bodies for deployment of PV modules. For this purpose, water surfaces such as lakes, dams, reservoirs, and sea are utilized to install PV panels. In floating PV systems, PV panels are mounted on pontoon-based floating structures that are attached to the bank or the water's bottom. The FPV systems have recently gained attraction due to the advancements in cost-effective floating solutions for large PV arrays. In addition, the lack of accessible land in island countries such as Japan, South Korea, and Singapore is driving this trend. In recent years, investment in FPV systems has increased, particularly

DOI: 10.1201/9781003272717-12

in India, which has a vast network of canals and reservoirs. Since this form of energy production is still in its infancy, it will require further study in future.

The construction and design of FPV systems are very important part for their durability and power generation. The deployment of PV systems on the water bodies is a bit tricky in comparison with land-based systems. The floating PV systems always have to experience various environmental loads; thus, the FPV system will have a different support structure. Floating structures are subjected to multiple loads such as wind load, dynamic water levels, wave motions, and water currents. These loads can change the orientation and tilt angle of the PV system and sometimes damage the floating structure. Therefore, these factors need to be carefully considered while designing the FPV system. The design and construction strategy of FPV is critical for its large-scale deployment and thus requires in-depth study into these elements. Furthermore, the performance and reliability of PV systems are influenced by the type of PV technology used and the local environment. Because of PV modules' closeness to water, the environmental conditions around the FPV systems differ from ground-based PV systems, and this might affect the degradation and performance of PV modules on the surface of water. The PV system reliability, planning, design, economic analysis, and cost-effective power generation all require long-term performance studies. Therefore, performance analysis and degradation aspects of FPV systems have been discussed in depth in this chapter.

The FPV can have good and bad environmental consequences, including affecting the amount of sunlight falling on water surface, oxygen entering the water, and wind energy mixing. The environmental and ecosystem impacts of FPV systems are poorly understood. However, recent research found that installing PV systems on water bodies had fewer environmental consequences than installing traditional PV systems. Deforestation during power plant implementation and bird death, runoff, erosion, site access, and microclimate change are all significant negative consequences of traditional ground-based PV systems. The PV systems built on water infrastructure, on the other hand, are able to reduce the problems encountered with traditional ground-based PV systems. However, it should be emphasized that covering water bodies with PV panels can reduce organic matter, which feeds fish and other aquatic species in the water, thereby disrupting the ecological balance. The man-made water bodies lack the original ecosystem of natural water bodies; therefore, the latter may get damaged more than the former. The PV panels above natural water bodies might alter the surrounding microclimate and cause local flora and fauna to suffer. As a result, man-made reservoirs, ponds, and lakes may be an appealing alternative for FPV system installation. Thus, the study of the environmental impact of FPV systems is also important for its widespread acceptance. The FPV system minimizes evaporation losses, which can save water and eventually be used for irrigation and drinking purposes. However, the design and structure of the FPV system have an impact on the evaporation reduction capability of the FPV system. The evaporation loss reduction is an important feature of FPV systems and needs to be studied properly for its accurate quantification.

12.2 FLOATING PV STATUS

Japan's National Institute of Advanced Industrial Science and Technology (AIST) developed the 20 kW capacity of first FPV plant and deployed it for research purposes in 2007 in Aichi, Japan [1]. The performance and water-cooling effect of Japan's first 20 kW FPV plant was studied by Ueda et al. [2]. For cooling the FPV module, intermittent watering was provided, and the water from the reservoir was used for intermittent cooling of modules. Due to the lower module temperature, the study found that the efficiency of the PV modules is improved. Following the success of this research effort, several FPV systems were deployed in countries such as India, China, Italy, Spain, France, South Korea, and the USA. From 2007 to 2013, Trapani and Redon Santafe [1] studied FPV installations around the world and recommended the use of thin-film PV module technology for vast lacustrine and maritime environments. Sahu et al. [3] observed a number of canal-top PV and FPV systems installed over the years and have pointed out numerous challenges and benefits associated with them.

FPV systems have widely been adopted around the world, with the current study estimating that FPV plants can satisfy roughly 25% of global energy demand while occupying only 1% of the available surface area over water bodies [4]. The most innovative and flexible FPV technologies are being developed by research centres all over the world. In this regard, Ferrer-Gisbert et al. [5] created a unique FPV system on a reservoir developed for agricultural purposes using a polyethylene-based floating module. From 2009 to 2014, Kim et al. [6] investigated several FPV systems in Korea and identified the huge potential in the country for large-scale FPV deployments.

Similarly, Hammoumi et al. [7] investigated the potential of Morocco for FPV deployment and assessed the efficiency of the floating PV system using a test bench. Additionally, research is being performed around the world to determine the viability of FPV systems in various operating conditions. To improve the operation of FPV systems on water bodies, Cazzaniga et al. [8] investigated the various installation designs and raft system. The authors have presented several FPV concepts that integrate concentrating, cooling, and tracking technologies in FPV plants. It must be noted that towards the end of 2019, the total global installed capacity of FPV systems had reached 2 GW, while by the end of August 2020, it had surpassed 2.6 GW [9–11]. However, the COVID-19 pandemic has had a significant impact on the expansion of the FPV industry, as on other industries. Looking at the data for the past 10 years, FPV systems have grown at a staggering rate of 133% each year, indicating that the technology will explode globally. However, given the rapid advancement of FPV technology over the span of last 5–10 years, a growth rate of 133% seems unlikely. Several constraints restrict the growth rate of FPV technology; as a result, according to the current research, the FPV technology is projected to expand at a rate of over 20% per year over the next 5 years [9].

The FPV technology is presently employed in over 35 countries, with over 350 systems in operation. The success of FPV deployment in these countries has inspired FPV deployment in other nations as well, with over 60 nations actively looking into it. Figure 12.1 depicts the global proportion of floating PV installations. Asia is the market leader in FPV, accounting for two-thirds of worldwide demand, with China,

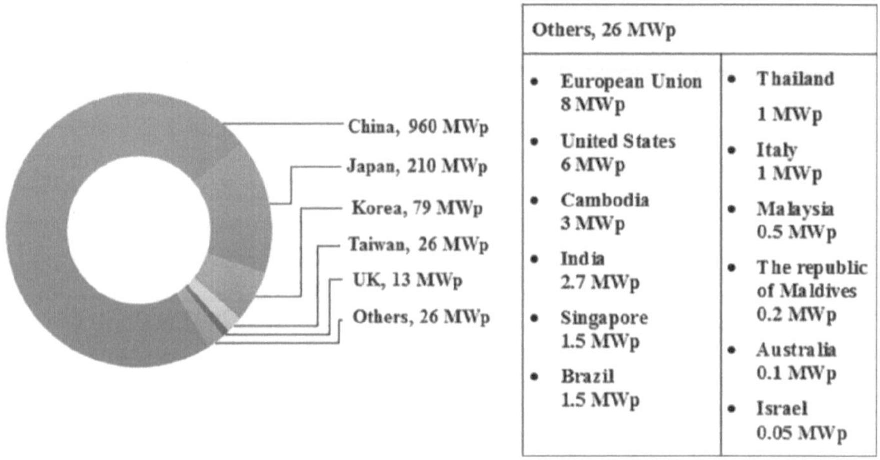

FIGURE 12.1 Global proportion of floating PV installations [10].

Vietnam, Thailand, Taiwan, South Korea, and India leading the way. China leads the FPV market and has already committed significant resources to it. Other Asian nations such as India, Taiwan, Korea, and Japan, on the other hand, are aggressively investing in this technology.

12.3 FPV SYSTEMS DESIGN AND STRUCTURE

FPV systems are comparable to land-based PV systems; however, the PV modules in FPV systems are mounted on a floating platform. An FPV system consists of PV arrays, inverters, metal frames, combiner boxes and lightning arresters and other components installed on a pontoon or separate float. Fibre-reinforced plastics (FRP) or high-density polyethylene (HDPE) are used to make these pontoons or floats. The pontoons or floats are held in place by anchoring and mooring systems, which prevent them from floating away. Figure 12.2 is a typical example of FPV installations.

12.3.1 Design and Structure

In typical FPV systems, multiple floating platforms are cascaded together to cover the surface of water bodies and provide foundation support for PV modules. While placing FPV systems on water bodies, several variables must be considered, including the ones listed below [13].

- Floating platform structure.
- In situ work for exploitation and construction.
- Reservoir's layout and varying water level.

The interior geometry and layout of reservoirs vary widely across the globe. Thus, it becomes difficult to adapt a universally deployable floating structure design [14].

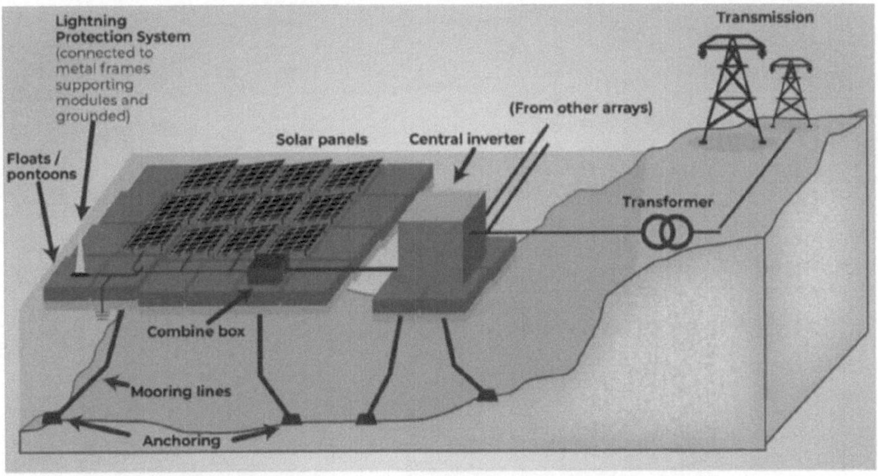

FIGURE 12.2 Typical example of FPV installations [10,12].

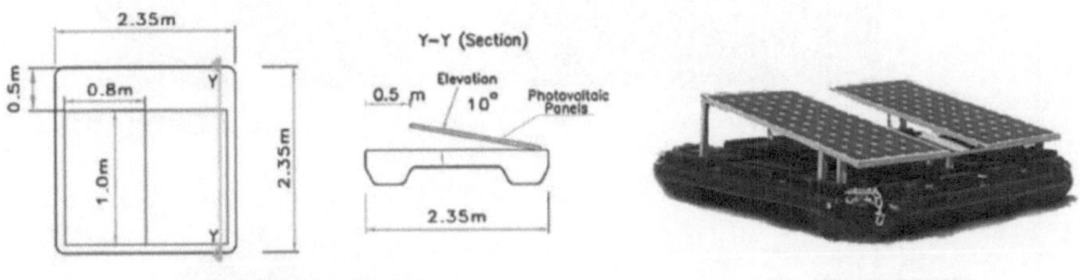

FIGURE 12.3 Typical design of FPV system [16,17].

The size of the floating modules is calculated by measuring the distance between frames. Herein, it is strongly recommended to have the access routes of at least 0.5 m to provide a walking area for easy operation and maintenance, as well as to reduce shadowing, as shown in Figure 12.3 [15].

Commercially available FPV systems are classified into three types based on their design and construction, and examples of these three types are given in Table 12.1 [18].

Type 1: Steel or aluminium rafts in combination with high-density polyethylene (HDPE) floating pipes.

Type 2: Both floating platform and rafts consisting of pure-floats design of HDPE.

Type 3: A floating pontoon is joined to form a huge platform that holds a PV module as a single unit.

Due to the mixed-type approach taken by various FPV developers, these three categories are occasionally comparable, making comparison difficult. Floating structure robustness, development cost, least water contamination and

TABLE 12.1
Floating PV Structures [18]

Type 1 Floating Structure

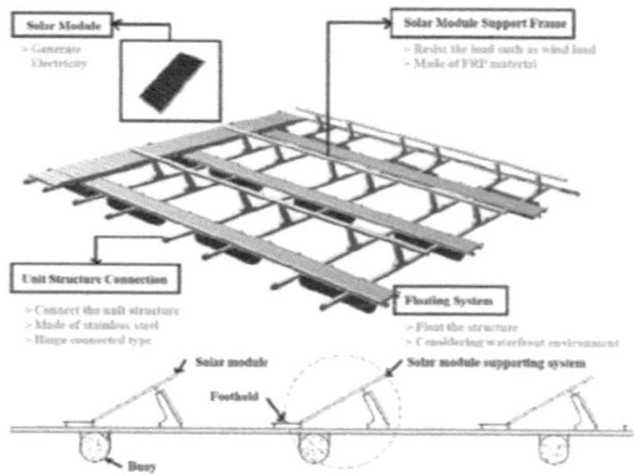

Type 2 Floating Structure

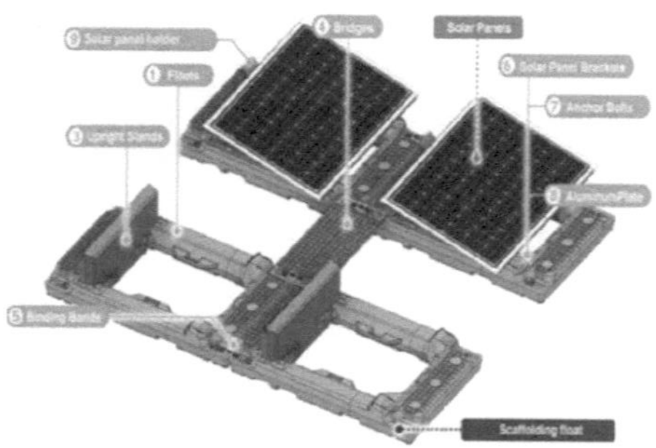

Type 3 Floating Structure

(Continued)

TABLE 12.1 (*Continued*)
Floating PV Structures [18]

environmental impact, ease in mooring and assembling, and flexibility of the system to adapt to local conditions are some of the major characteristics to consider when analysing the benefits of various design methodologies (tracking, cooling, panel tilt, etc.). It's impossible to cover every FPV structure and design because the market of FPV is quickly growing and new innovative concepts are constantly emerging. As a result, this chapter focuses solely on the most prevalent and popularly used FPV structures.

12.4 FPV SYSTEM'S PERFORMANCE AND DEGRADATION ASPECTS

The degradation in PV modules have a substantial impact on performance and the amount of energy a PV module generates in any given climate. These PV module or system parameters are determined by the characteristics of PV technology used and the local climatic conditions in which the PV system is installed. The PV arrays installed in floating PV systems are mounted on water bodies, where the local environmental conditions differ from those on land. Because of evaporation from aquatic bodies, the environment is more humid and cold than it is on land. Floating PV systems will perform and degrade differently from land-based PV systems due to excess humid and cool conditions [19,20]. This section will look at the performance and degradation aspects of FPV systems to get a better understanding of how they behave at the water's surface. Furthermore, because each PV technology reacts differently to various climatic circumstances, the type of PV technology has a significant impact on PV system performance [11,19]. As a result, the performance of several PV module technologies under humid climatic conditions of water bodies has been investigated in order to determine which PV module technology is ideal for large-scale FPV system installation.

12.4.1 PERFORMANCE ANALYSIS

In order to ensure their long-term reliability and quality, PV modules are subjected to a number of performance requirements and qualification tests in accordance with the International Electrotechnical Commission (IEC) standards [21–25]. PV module performance is known to degrade in real-world outdoor circumstances since PV modules' output changes with local climatic conditions [26–29]. As a result, performance analysis is a critical component of PV system planning, sizing, and installation. Previous research on the performance of PV modules depending on the materials used in their manufacturing, geographical location, temperature coefficient, and solar radiation has given a good amount of information. IEC 61724 provides a number of performance indicators for assessing the performance of PV systems or technologies in different climates. One of the most commonly utilized performance indicators to identify a system's genuine efficiency is the performance ratio (PR), which is the ratio of the final yield to the reference yield. This enables for a more accurate comparison of different PV module technologies deployed in the same environment as well as PV systems put in various places.

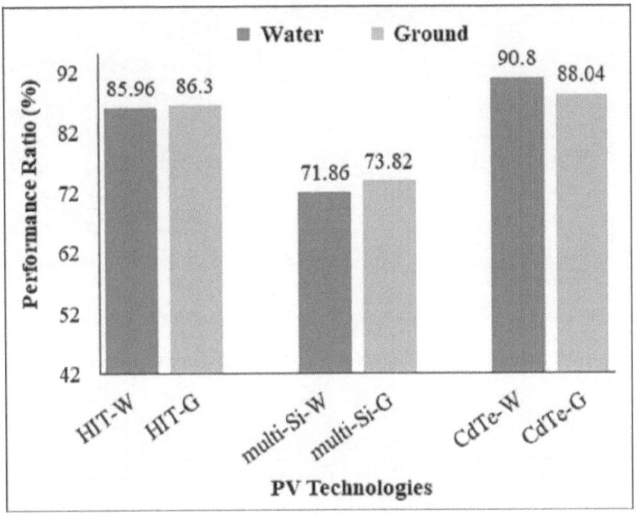

FIGURE 12.4 Performance analysis of different PV technologies on water and land surfaces [30].

Different PV technologies on the water surface have recently been investigated and compared to equivalent PV technologies on the land surface, as illustrated in Figure 12.4 [30]. On the water surface, the performance of c-Si-based technologies such as HIT and multi-Si modules is lower than that of equivalent PV systems on land. A thin-film technology such as CdTe, on the other hand, performs better on the water surface than the equivalent CdTe technology on the land surface. Despite having a low module temperature over water bodies, the performance of c-Si-based PV systems on the water surface is proven to be poorer. The high humidity found in water bodies has a deleterious influence on c-Si modules, rather than benefiting from the low module temperature.

The output power of a PV system is clearly reliant on the characteristics of PV technology used and the local environmental circumstances, as shown in the above analysis. In hot and humid climates, the thin-film CdTe technology has also been demonstrated to be superior to c-Si technology. The spectral response of PV technologies has a considerable impact on their performance at a certain site. The thin-film technology such as CdTe captures only the blueshift spectrum (visible spectrum) of solar radiation for energy conversion, but c-Si-based technologies use both the blueshift (visible spectrum) and the redshift (infrared spectrum) [31,32]. The spectral response of the thin-film technology is in the range of 300–900nm, while that of c-Si technology is 300–1100nm. As seen in Figure 12.5, water vapour absorbs the infrared part of the solar spectrum, but not the visible part (b). As a result of these observations, the humid atmosphere has a detrimental impact on solar radiation's redshift spectrum, but has no effect on the blueshift spectrum [31]. As a result, the performance of c-Si modules under humid conditions is shown to be lower than that of thin-film modules. Modules in floating PV systems are always exposed to a very wet atmosphere due to the evaporation

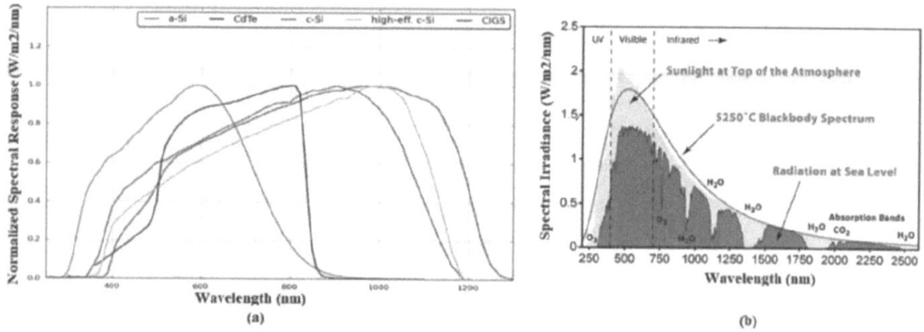

FIGURE 12.5 (a) PV technologies' spectral response. (b) Solar radiation's spectrum [31,33].

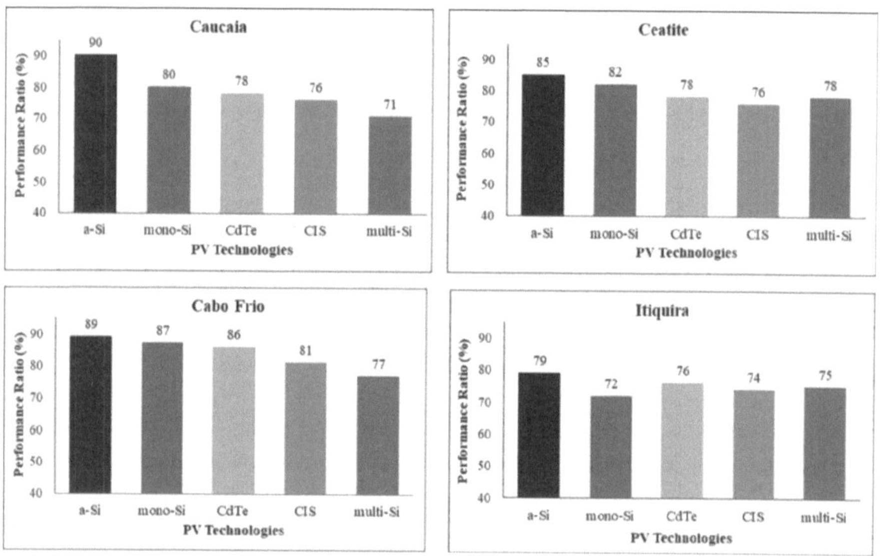

FIGURE 12.6 Performance of different PV technologies in various Brazilian climates [34].

process, which may have an impact on the performance of c-Si-based PV systems, one of the widely deployed PV technologies. As a result, the performance of various PV technologies on the water surface should be investigated before going forward with installation of FPV on a large scale.

Several investigations on the performance of various PV technologies in various environmental situations have previously been conducted. Figure 12.6 shows how different PV technologies perform in different climates in Brazil [34]. PV technology's performance is determined by its location. Temperature, solar insolation, and humidity are only a few of the primary environmental elements that influence the effectiveness of PV technology. Thin-film technologies (a-Si and CdTe) have low temperature coefficients and are thus less impacted by high temperatures, but c-Si-based technologies, both multi-Si and mono-Si, have large temperature coefficients and hence are substantially influenced by high temperatures. However, the a-Si technology is most suited to hot climates as it has the lowest temperature coefficient of all the approaches studied. Furthermore, PV technologies' spectral response influences their performance at a given location, and thin-film technologies outperform in humid environments as they have a blueshift spectrum. As illustrated in Figure 12.6, thin-film technologies such as a-Si and CdTe outperform multi-Si PV systems. The low temperature coefficient and blueshift spectrum response of the thin-film technology help to achieve a high performance.

12.4.2 Degradation Analysis

The commercialization of PV systems for power generation requires the long-term reliability of PV modules. PV system's investment returns and cost-effective power production are affected by many variables, one of which is performance degradation due to environmental conditions. The IEC standards recommend performing indoor accelerated stress testing to better understand degradation processes and PV module certification under specific conditions. The goal of these indoor experiments is to simulate the degradation mechanism that occurs in a real-world environment. The standardized methods aided in ensuring and promoting PV module durability and reliability. Internal accelerated stress tests, on the other hand, are unable to duplicate all degradation processes because a few degradation modes are only visible after substantial years of field exposure. Actual outdoor conditions have a considerable impact on degradation modes, which aren't confined to the manufacturer-specified performance degradations.

The reliability of PV modules is affected by these degradation mechanisms, which have major implications on power generation. Despite recent improvements in PV module quality and reliability, the modules are still susceptible to degradation when exposed to various environmental stressors in the field. The primary environmental stressors that can impact the PV module's reliability are as follows [35–37]:

Ultraviolet (UV) radiation
High temperature and thermal cycling
Humidity.

Due to thermomechanical, structural, and chemical changes, these environmental stressors can affect the performance and longevity of PV modules. PV modules on the water are constantly exposed to a humid environment, and the module temperature drop is minimal; some studies reported a 2–5°C fall in module temperature [11,30,38]. As a result, environmental stresses including temperature, ultraviolet light, and humidity can have negative effects on modules in water bodies. From Figure 12.7, one can see that PV systems can suffer from a variety of degradation modes.

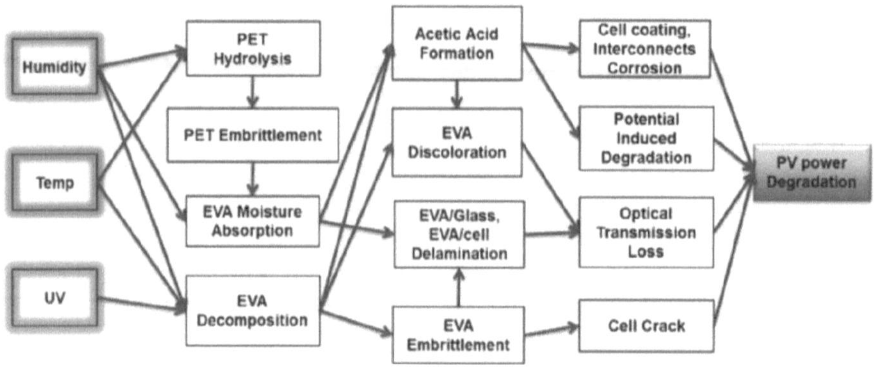

FIGURE 12.7 Degradation modes in PV systems due to a combination of primary environmental stressors [39].

The PV module degradation must be identified and analysed, which aids in the progress of PV technologies and guarantees that PV systems are durable and reliable. To investigate the degradation modes in PV modules, various approaches such as visual inspection, laser beam-induced current (LBIC), I–V characteristics, ultrasonic inspection, imaging, infrared imaging (IR imaging), electroluminescence (EL) imaging, and dark lock-in thermography (DLIT) can be used.

12.5 EVAPORATION IN FPV SYSTEMS

There is great potential in the implementation of PV systems on existing water bodies. According to the World Bank report, there is approximately 4 lakh km^2 of water surface area globally, a part of which can be used for FPV systems [40,41]. In addition, evaporation minimization is one of the key aspects of FPV systems, among other possible benefits. Water is an economic good, according to the UN's Dublin Principle (1992), and therefore can be assigned a monetary value [42]. The monetary value of water can be understood using the example of Indian state of Maharashtra, which has set multiple rates for water, with the lowest charge for flow irrigation being 119 INR/ha and the highest fee being 6297 INR/ha, depending on the irrigation pattern [42]. As a result, calculating the quantity of water saved by FPV systems is crucial for establishing their economic viability and life cycle cost analysis. Figure 12.8 depicts how the evaporation reduction capabilities of FPV systems is affected by different types of FPV plants and installation topologies.

Figure 12.8a shows a floating PV system that consists of a modular floating platform covering the whole surface beneath the module and virtually eliminates solar radiation transmission to the water surface. The evaporation may be reduced by 49% with this type of typology [43]. The floating systems and canal-top installations with modules coupled to a tubular buoyancy system, as shown in Figure 12.8b and c, provide adequate ventilation and evaporative cooling of the modules. The water surface area in these types of installations are not fully covered; therefore, solar radiation cannot be totally blocked.

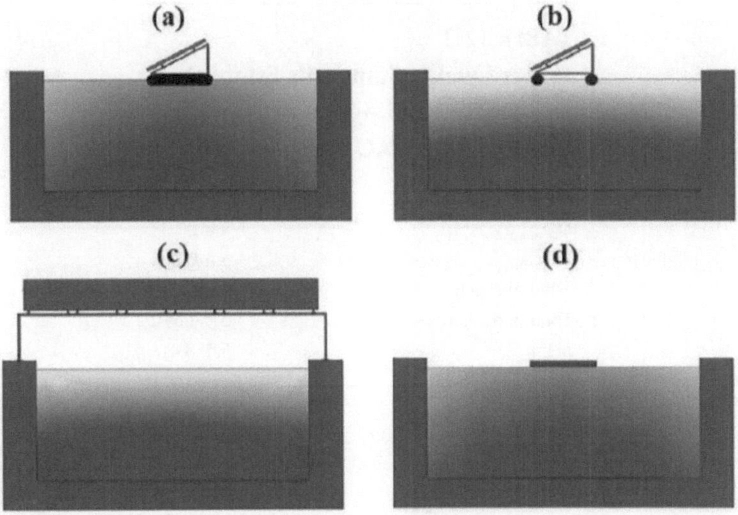

FIGURE 12.8 FPV system installation typologies: (a) floating platform that covers the whole surface beneath the module, (b) modules coupled to a tubular buoyancy system, (c) canal-top PV systems, and (d) flexible floats [43].

These devices only block direct solar radiation while allowing diffuse solar energy to get through. The evaporation may be reduced by 30% with this type of typology [31].

Furthermore, flexible floats, as shown in Figure 12.8d, are submerged PV modules with direct water contact; and with this type of typology, evaporation can be reduced by up to 42% [43]. Kumar et al. [11] established an analytical methodology to evaluate the reduction in evaporation in floating and canal-top PV systems (see Figure 12.8b and c). They developed a model and discovered that evaporation was reduced by 36%. As the FPV systems are often built on a portion of a total water surface, the exact quantification of reduction in evaporation loss is currently challenging. As a result, further research is required to support the assertions made above. Also, covering water bodies with PV modules might diminish organic matter, which feeds fish and other aquatic creatures, thereby disrupting the ecological balance. However, man-made water bodies lack the same basic ecosystem as natural water bodies, so natural water bodies may be more affected.

12.6 FLOATING PV ENVIRONMENTAL IMPACTS

Very few researchers have studied the environmental impacts of floating PV systems. Much of the research in this area has focused on technology improvement rather than environmental consequences. There is very little research in this area, and most of it is theoretical [44]. A recent study looked into the impact of FPV systems on the quality of water in the water bodies having FPV system installed over them. Using a computer model of the 50 km long canal channel in Egypt, the scientists examined numerous water quality indicators such as dissolved oxygen

TABLE 12.2
Water Quality Standards Established by the Food and Agriculture Organization (FAO) [45]

Parameter	Value
$DO_{irrigation}$	4 mg/L
$DO_{freshwater}$	5 mg/L
Total nitrogen	5 mg/L
Total phosphorus	0–2 mg/L
pH	6.5–8.4

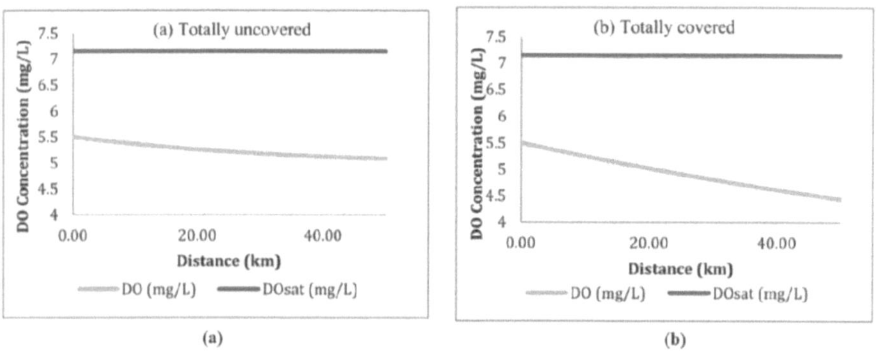

FIGURE 12.9 DO concentration in the (a) completely uncovered canal and (b) completely covered canal [44].

(DO), alkalinity, compounds – pH, algae, and nutrients – including phosphate and nitrogen.

Table 12.2 summarizes the water quality standards established by the Food and Agriculture Organization (FAO).

The dissolved oxygen (DO) content of a completely covered 50 km long canal was found to be lower than that of an uncovered canal. The DO is a critical water variable that affects the chemical, physical, and biological processes in bodies of water. Figure 12.9a and b shows how the DO concentration varies over the canal's length (b). The DO concentration at the canal end of a completely covered canal is 4.47 mg/L, whereas the DO concentration at the canal end of a completely uncovered canal is 5.09 mg/L. Similar findings of reduced DO content in the water under floating PV systems have been reported by Ziar et al. [46]. As a result, the PV cover reduces DO levels in water bodies, resulting in hypoxia beneath floating and canal-top PV systems.

Incoming solar light is blocked by the floating and canal-top PV systems, which has a direct impact on the photosynthesis process and, as a result, has a major impact

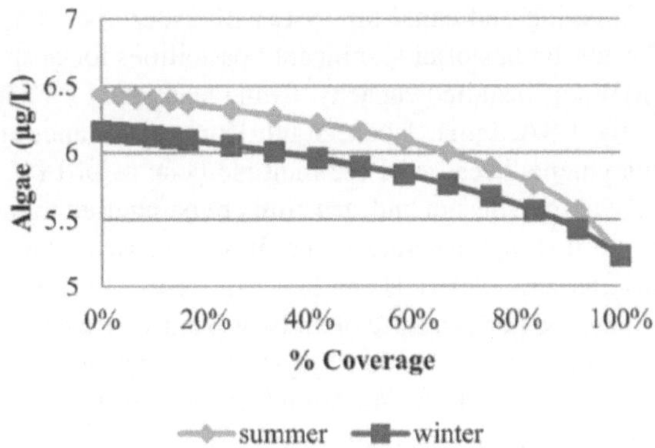

FIGURE 12.10 Algae concentration in the canal water [44].

on the concentration of algae in water bodies. Figure 12.10 demonstrates that the algal concentration in a completely exposed canal is 6.44 g/L in the summer and 6.16 g/L in the winter, whereas it is 5.23 g/L in both the winter and summer seasons in a completely covered canal.

Similarly, Ziar et al. [46] saw a decrease in algae and plant growth beneath the FPV systems as well. As a result, it's reasonable to conclude that floating and canal-top PV systems lead to reduction in algae levels in water bodies. Variations in the wind around the modules, on the other hand, may have an impact on sediment formation processes; therefore, increases in microbial activity cannot be neglected. As a result, regardless of the cause of increased hypoxia under FPV systems, the environmental impact of FPVs (e.g., greenhouse gas emissions, impact on fish, and internal loads) can be far-reaching [46].

The organic nitrogen concentration is slightly reduced as a result of PV cover. Similarly, PV cover has little effect on nitrate nitrogen, which falls to some extent. Ammonia nitrogen, on the other hand, reacts differently under PV cover, increasing slightly. Furthermore, organic and inorganic phosphorus levels in water were shown to be lower under the floating PV system, while the reduction was minor. The pH is influenced by alkalinity, whereas alkalinity is influenced by chemical and biological processes. Due to the covering of water bodies by FPV systems, some processes will be inhibited, while others will be accelerated, resulting in increased level of alkalinity and, eventually, a rise in pH [44]. More research is needed to determine the exact proportion of these factors influenced by floating PV systems. As a result, PV cover has a significant impact on DO and algal concentration. This conclusion, however, is based on very little study. More research is needed to understand the impact of floating and canal-top PV systems on local ecosystem.

12.7 CONCLUSIONS

Floating photovoltaic (FPV) systems have the potential to be a significant contributor to the global development of renewable energy sources. There is a lot of room for the installation of floating and canal-top systems because the FPV system does not require land and water bodies offer significant possibilities for hosting such systems. With over 1000 MWp of installed capacity, China leads the FPV industry. However, countries such as the USA, India, Japan, South Korea, Thailand, and Vietnam have boosted their deployments because of the multiple benefits of FPV. Currently, three types of floating PV system design and structure are being used to install FPV plants. In FPV systems, the floating structure is the most expensive component. However, innovative designs and new materials can lower this cost. Since water bodies have a highly humid climate, the performance of an FPV plant varies depending on the type of PV technology used. The c-Si-based PV systems are being used worldwide due to their low cost and wide availability. In humid environments, c-Si PV technologies perform poorly, but thin-film PV technologies do well. As a result, choosing the right PV technology is critical for large-scale FPV plant deployment because FPV systems are constantly exposed to high humidity levels. In addition, the highly humid environment also negatively affects the lifetime and reliability of the PV module. Thus, long-term degradation studies of FPV systems are needed to ensure their reliability. Furthermore, evaporation reduction is a key characteristic of FPV systems, and the degree of evaporation reduction varies according to the type of FPV plant structure and design. When the FPV system covers the entire water surface, evaporation can be decreased by up to 49%, while modules deployed at a predetermined height to enable diffuse radiation on the water surface can reduce evaporation by 30%. The environmental impact of FPV systems is not fully understood. Despite the fact that the majority of FPV systems are built on man-made water bodies, they can have a substantial impact on water ecology. The environmental impact of FPV systems is not fully understood. The amount of PV cover has a big impact on water quality factors such as dissolved oxygen and algal concentration. The FPV system has an impact on other variables including as pH, alkalinity, and nutrients. However, more research is needed to determine the environmental impact of FPV systems.

ACKNOWLEDGEMENTS

The authors would like thank the Indian Institute of Technology Bombay, Mumbai, India, for providing the resources and financial support to carry out this research.

ABBREVIATIONS

a-Si	Amorphous silicon
CdTe	Cadmium telluride
c-Si	Crystalline silicon
DLIT	Dark lock-in thermography
DO	Dissolved oxygen
EL	Electroluminescence
FAO	Food and Agriculture Organization
FPV	Floating photovoltaic
FRP	Fibre-reinforced plastics
HDPE	High-density polyethylene

IEC International Electrotechnical Commission
IR Infrared
LBIC Laser beam-induced current
PR Performance ratio
PV Photovoltaic
UV Ultraviolet

REFERENCES

[1] Trapani K, Redón Santafé M. A review of floating photovoltaic installations: 2007–2013. *Progress in Photovoltaics: Research and Applications*, 2015;23:524–32.

[2] Ueda Y, Sakurai T, Tatebe S, Itoh A, Kurokawa K. Performance analysis of PV systems on the water. *23rd European Photovoltaic Solar Energy Conference*, 2008:2670–3.

[3] Sahu A, Yadav N, Sudhakar K. Floating photovoltaic power plant: A review. *Renewable and Sustainable Energy Reviews*, 2016;66:815–24.

[4] Marco TG, Raniero C, Marco R-C, Paolo R-C. Geographic and technical floating photovoltaic potential. *Thermal Science*, 2018;22:831–41.

[5] Ferrer-Gisbert C, Ferrán-Gozálvez JJ, Redón-Santafé M, Ferrer-Gisbert P, Sánchez-Romero FJ, Torregrosa-Soler JB. A new photovoltaic floating cover system for water reservoirs. *Renewable Energy*, 2013;60:63–70.

[6] Kim S-H, Yoon S-J, Choi W, Choi K-B. Application of floating photovoltaic energy generation systems in South Korea. *Sustainability*, 2016;8:1333.

[7] El Hammoumi A, Chalh A, Allouhi A, Motahhir S, El Ghzizal A, Derouich A. Design and construction of a test bench to investigate the potential of floating PV systems. *Journal of Cleaner Production*, 2021;278:123917.

[8] Cazzaniga R, Cicu M, Rosa-Clot M, Rosa-Clot P, Tina G, Ventura C. Floating photovoltaic plants: Performance analysis and design solutions. *Renewable and Sustainable Energy Reviews*, 2018;81:1730–41.

[9] Haugwitz F. Floating solar PV gains global momentum. *PV Magazine International – Photovoltaic Markets and Technology*, 2020. https://www.pv-magazine.com/2020/09/22/floating-solar-pv-gains-global-momentum/ (accessed on 25 May, 2021).

[10] Acharya M, Devraj S. *Floating Solar Photovoltaic (FSPV): A Third Pillar to Solar PV Sector?* The Energy and Resources Institute (TERI), New Delhi, 2019.

[11] Kumar M, Kumar A, Gupta R. Comparative degradation analysis of different photovoltaic technologies on experimentally simulated water bodies and estimation of evaporation loss reduction. *Progress in Photovoltaics: Research and Applications* 2020;29(3):357–78.

[12] Gorjian S, Sharon H, Ebadi H, Kant K, Scavo FB, Tina GM. Recent technical advancements, economics and environmental impacts of floating photovoltaic solar energy conversion systems. *Journal of Cleaner Production* 2020:124285.

[13] Santafé MR, Soler JBT, Romero FJS, Gisbert PSF, Gozálvez JJF, Gisbert CMF. Theoretical and experimental analysis of a floating photovoltaic cover for water irrigation reservoirs. *Energy* 2014;67:246–55.

[14] Siecker J, Kusakana K, Numbi B. A review of solar photovoltaic systems cooling technologies. *Renewable and Sustainable Energy Reviews* 2017;79:192–203.

[15] Redón-Santafé M, Ferrer-Gisbert P-S, Sánchez-Romero F-J, Torregrosa Soler JB, Ferran Gozalvez JJ, Ferrer Gisbert CM. Implementation of a photovoltaic floating cover for irrigation reservoirs. *Journal of Cleaner Production* 2014;66:568–70.

[16] Trapani K, Redón Santafé M. A review of floating photovoltaic installations: 2007–2013. *Progress in Photovoltaics: Research and Applications* 2015;23:524–32.

[17] Ferrer-Gisbert C, Ferrán-Gozálvez JJ, Redón-Santafé M, Ferrer-Gisbert P, Sánchez-Romero FJ, Torregrosa-Soler JB. A new photovoltaic floating cover system for water reservoirs. *Renewable Energy* 2013;60:63–70.
[18] Cazzaniga R. *Floating PV Structures. Floating PV Plants*, Elsevier. 2020, 33–45.
[19] Kumar M, Kumar A. Performance assessment of different photovoltaic technologies for canal-top and reservoir applications in subtropical humid climate. *IEEE Journal of Photovoltaics* 2019;9:722–32.
[20] Kumar M, Chandel S, Kumar A. Performance analysis of a 10 MWp utility scale grid-connected canal-top photovoltaic power plant under Indian climatic conditions. *Energy* 2020:117903.
[21] Commission IE. *Photovoltaic System Performance Monitoring-Guidelines for Measurement, Data Exchange and Analysis*, IEC. 1998, 61724.
[22] Commission IE. *Thin Film Terrestrial Photovoltaic (PV) Modules – Design Qualification and Type Approval*, IEC International Standard, Geneva, Switzerland, Tech Rep. 2008, 61646.
[23] Jamaly M, Bosch JL, Kleissl J, Zheng Y. Performance analysis of power output of photovoltaic systems in San Diego County. *2012 IEEE Power and Energy Society General Meeting*, IEEE. 2012, 1–7.
[24] Photovoltaic CST. *Modules—Design Qualification and Type Approval*, IEC. 2005, 1215, 2005.
[25] Sharma V, Chandel S. Performance and degradation analysis for long term reliability of solar photovoltaic systems: A review. *Renewable and Sustainable Energy Reviews*, 2013;27:753–67.
[26] Colli A, Sparber W, Armani M, Kofler B, Maturi L. Performance monitoring of different PV technologies at a PV field in Northern Italy. *25th European Photovoltaic Solar Energy Conference and Exhibition/5th World Conference on Photovoltaic Energy Conversion*, 2010:6–10.
[27] Dash P, Gupta N, Rani P, Chakraborty S, Sadhu P. Techno-economic performance analysis of four photovoltaic technology at North-East Region of India. *2018 International Conference on Power Energy, Environment and Intelligent Control (PEEIC)*, IEEE. 2018, 751–4.
[28] Perveen G, Rizwan M, Goel N. Comparison of intelligent modelling techniques for forecasting solar energy and its application in solar PV based energy system. *Energy Systems Integration*, 2019;1:34–51.
[29] Rizwan M, Chaudhary P, Ahmad T. Performance analysis of maximum power point tracking techniques for photovoltaic systems. *Advanced Science Letters*, 2014;20:1231–47.
[30] Kumar M, Kumar A. Experimental characterization of the performance of different photovoltaic technologies on water bodies. *Progress in Photovoltaics: Research and Applications*, 2020;28:25–48.
[31] Dirnberger D, Blackburn G, Müller B, Reise C. On the impact of solar spectral irradiance on the yield of different PV technologies. *Solar Energy Materials and Solar Cells*, 2015;132:431–42.
[32] Rüther R, Livingstone J. Seasonal variations in amorphous silicon solar module outputs and thin film characteristics. *Solar Energy Materials and Solar Cells*, 1995;36:29–43.
[33] *Solar Radiation Spectrum. Solar Spectrum.* https://en.wikipedia.org/wiki/Sunlight#/media/File:Solar_spectrum_en.svg (accessed on 3 June, 2021).
[34] do Nascimento LR, Braga M, Campos RA, Naspolini HF, Rüther R. Performance assessment of solar photovoltaic technologies under different climatic conditions in Brazil. *Renewable Energy*, 2020;146:1070–82.

[35] Gabor AM, Janoch R, Anselmo A, Lincoln JL, Seigneur H, Honeker C. Mechanical load testing of solar panels – Beyond certification testing. *2016 IEEE 43rd Photovoltaic Specialists Conference (PVSC)*, IEEE. 2016, 3574–9.

[36] Santhakumari M, Sagar N. A review of the environmental factors degrading the performance of silicon wafer-based photovoltaic modules: Failure detection methods and essential mitigation techniques. *Renewable and Sustainable Energy Reviews*, 2019;110:83–100.

[37] Suleske A, Singh J, Kuitche J, Tamizh-Mani G. Performance degradation of grid-tied photovoltaic modules in a hot-dry climatic condition. *Reliability of Photovoltaic Cells, Modules, Components, and Systems IV*, International Society for Optics and Photonics. 2011, 81120P.

[38] Liu H, Krishna V, Lun Leung J, Reindl T, Zhao L. Field experience and performance analysis of floating PV technologies in the tropics. *Progress in Photovoltaics: Research and Applications*, 2018;26:957–67.

[39] Rosa-Clot M, Tina GM. *Cooling Systems. In Floating PV Plants*, Academic Press. 2020, 67–77.

[40] *Where Sun Meets Water: Floating Solar Market Report*, World Bank Group and SERIS, Singapore. 2018.

[41] Kumar M, Kumar A. Experimental validation of performance and degradation study of canal-top photovoltaic system. *Applied Energy*, 2019;243:102–18.

[42] Sarkar SK. It's time to overhaul water pricing norms. *The Hindu BusinessLine*, 2019. https://www.thehindubusinessline.com/opinion/its-time-to-overhaul-water-pricing-norms/article27071838.ece (accessed on 30 May 2021).

[43] Scavo FB, Tina GM, Gagliano A, Nizetic S. An assessment study of evaporation rate models on a water basin with floating photovoltaic plants. *International Journal of Energy Research*, 2021;45:167–88.

[44] Baradei SE, Sadeq MA. Effect of solar canals on evaporation, water quality, and power production: An optimization study. *Water*, 2020;12:2103.

[45] Ayers RS, Westcot DW. *Water Quality for Agriculture*, Food and Agriculture Organization of the United Nations Rome. 1985.

[46] Ziar H, Prudon B, Lin FY, Roeffen B, Heijkoop D, Stark T, et al. Innovative floating bifacial photovoltaic solutions for inland water areas. *Progress in Photovoltaics: Research and Applications*, 2020;29(7):725–43.

13 Waste Heat Recovery Technologies for Sustainability and Economic Growth in Developing Countries

Sunday O. Oyedepo and Adebayo B. Fakeye
Covenant University

CONTENTS

13.1 INTRODUCTION

Energy is a must-satisfied need of life that is typically sought based on the economic power of an individual, industry, or nation. The quality and quantity of the energy consumption indicate the quality of life of the citizens and the level of industrial, technological, and socio-economic development of the nation. Since the discovery of fossil fuels, industrial growth and economic prosperity of any country are affiliated

DOI: 10.1201/9781003272717-13

with the consumption of resources (Lee et al., 2015). Fossil fuels account for more than 80% of the growing world energy consumption. Still, the thermal conversion of these limited resources is shallow, typically less than 0.35, which means more than 65% of the energy contents of the fuels are wasted away as low-grade heat. This constitutes thermal pollution, excessive release of greenhouse and other toxic gases into the atmosphere, and inefficient deployment and overexploitation of the finite resources, resulting in their fast depletion.

Hence, global attention on the associated hazards gave rise to the call for the energy transition to renewable energy resources. However, despite the consent of many countries to the Kyoto Protocol and subsequent Accords on global warming and energy transition, reports show that the current energy mix is still largely reliant on fossil fuels. Although the capacity of RE is gradually increasing, its proportion in the energy mix is on a steep decline (BP Statistical Review of World Energy, 2019). National economic development worldwide is entirely dependent on the conventional energy source. According to a report by the International Energy Agency (IEA), based on present trends in energy use and efficiency, the temperature rise is anticipated to reach 6°C by 2050, with disastrous implications (Rahbar et al., 2017). A subsequent review by British Petroleum (BP) in 2018 indicated that the global energy demand and the consequential carbon emissions increased at their most excellent rates since 2010, defying the accelerated transition predicted by the Paris climate goals. According to BP Statistical Review of World Energy (2019), decarbonizing the power industry while meeting the fast-growing energy demand, mainly in developing countries, is the only and most significant difficulty confronting the global energy system over the next 20 years. While also acknowledging the importance of RE deployment in overcoming environmental impacts of energy use, it was factually admitted that RE could not meet the challenges.

Miró et al. (2016) opined that most developing nations are confronted with the discharge of a massive amount of usable low-grade waste heat to the environment with complete recovery and reuse. Waste heat is recognized as a valuable resource for sustainable development. However, the resource stays widely unutilized, but discharged into the environment. Similarly, many recent reviews and assessments on energy conducted by global authorities have identified energy efficiency as a priority in the energy policy agenda capable of achieving global climate goals, increasing the energy security of national economies, and providing jobs (Rocco, Ferrer, and Colombo, 2018). The two factors responsible for the increase in demand for final energy in developing nations are rapid growth in population and the drive to increase their industrial base. Although the thermal conversion of fossil fuels has typically low conversion efficiencies of 30%, industrialization in growing economies is mostly driven by fossil fuels primarily due to their economic advantage, especially coal, which is relatively abundant and cheaper. The same applies to the power sector, though with better CO_2 mitigation potential, because its emissions are localized and easy to control than the industrial sector with distributed emissions (Chen et al., 2018).

Hence, several works have been carried out, and many still ongoing waste heat recovery technologies mitigate the unsustainable utilization of fossil fuels and reduce greenhouse gas (GHG) emissions. Cantore, Calì and Velde (2016) examined the impact of energy efficiency (EE) on modern energy conversion devices and economic growth in response to the conflict between economic and environmental policies in developing nations working towards industrialization. They found out that lower energy intensities corresponded to higher total factor productivity for nearly all the countries at micro- and macro-levels. Their result, in essence, implies that increasing the energy efficiency of developing economies will improve economic growth. Marzi et al. (2019) identified that the lack of standard indices to compare energy consumption and power production across countries is a major barrier to enhancing energy efficiency. Hence, they developed an index for developing countries composed of 30 indicators under political, economic, social, and technological factors to appraise their energy efficiency potentials and associated investment risks to encourage sustainable use of fossil fuels.

Muhammad (2019) investigated the relationships between energy consumption, carbon footprints, and economic growth and established that economic growth increases with energy consumption. Their recommendation favored the adoption of sustainable technologies to decrease CO_2 emissions. In setting the goals explicitly personalized to promoting sustainable energy in developing countries, Lee, Leal and Dias (2019) proposed six plans, namely enhancing primary energy security and reliability of the final energy system, cost reduction (capital, operation, and maintenance), the global effect of the energy system on climate, and the environmental impact of the energy system on the locality. Considering the present increase in the growth rate in population, specifically in non-industrial nations, combined with the exhaustion of fossil fuel sources, waste heat recovery devices present an appealing choice that advances energy conservation and preservation of natural resources from additional degradation. Moreover, waste heat recovery offers an impressive decrease in the environmental impact of fossil fuel combustion.

Unlike material waste, which is easy to detect and quantify in terms of quantity and quality, waste heat is more complex to identify and quantify. As a result, understanding waste heat and viable solutions for collecting it offers a once-in-a-lifetime chance to reduce manufacturing and process costs while also lowering environmental impacts. Using low-quality energy from exhaust heat sources is a must if you want to improve energy conversion system performance while lowering pollution. There have been a few developments that have the potential to convert waste heat into usable products: System types include thermal, solid state (thermoelectric generators and piezoelectric), and hybrid.

The principal aims of this chapter are to (i) evaluate waste heat resources and ascertain the potential of waste heat recovery (WHR) technologies as a pathway to sustainability and socio-economic growth in developing countries; (ii) assess the potential of WHR systems as sustainable technologies to proffer solution to the challenges of environmental pollution due to combustion of fossil fuel, and (iii) discuss the benefits and barriers to the implementation of WHR technologies in developing countries.

13.2 WASTE HEAT RESOURCES AND POTENTIALS

Waste heat is latent heat that is dissipated to the environment from an industrial process at a high temperature above the production plant's ambient temperature.

A reasonable quantity of the waste heat source can be cheaply recovered and utilized (Upathumchard, 2014; Oyedepo and Fakeye, 2020). Combustion processes in industrial operations, power plants, motor vehicles, and other sources of waste heat generation are the primary sources of waste heat generation. Low-grade waste heat can be released into the environment as a result of poor maintenance and low-grade equipment in an industrial process (Elson and Hampson, 2015; Kandathil, 2016).

Recoverable waste heat can be assessed based on quality, and this can be classified based on temperature range such as high quality (650°C), medium quality (232–649°C), and low quality (232°C and below) (Fakeye and Oyedepo, 2019). We have different approaches used to assess the waste heat potential: the theoretical (or physical) potential, the technical potential, and the economically feasible potential (Brückner et al., 2015; Elson and Hampson, 2015; Singh and Pedersen, 2016).

The theoretical potential estimates the heat quality with temperature over the surrounding temperature and can be harvested from the heat source. In contrast, the technical potential assesses the lowest permissible temperature of removing and utilizing low-grade heat and possible technology to adopt. The potential economic approach emphasizes economic implications, financial parameters, and technical potential of harvesting low-grade heat sources. It is essential to consider the proper technique for the selection of the appropriate WHR method. It has been established that the possibility of recovery of waste heat from its source ranges between 32% and 80% depending on the temperature level when the temperature is more than 150°C (Kurle et al., 2016). Moreover, the work of Brueckner et al. (2014) reported that the potential of waste heat recoverable from dissipated heat in the European industrial sector varied from 5% and 30% of the energy used of the sector and recoverable waste heat potential from the sector was put at 370.41 TWh (Panayiotou et al. 2017).

13.2.1 Potential Waste Heat Recovery Techniques

The low-temperature heat dissipated from industrial processes and power plants operation is unavoidable. However, the various low-temperature heat streams scattered to the environment can be utilized as a source for heating, cooling, or power generation systems. The techniques for low-grade heat usage in the industry are classified as either passive or active. Figure 13.1 depicts the different classifications of technologies depending on whether heat is immediately recycled at the same or lower temperature, transformed to another source of energy, or upgraded to a higher temperature (Brückner et al., 2015). The focus of this chapter is on waste heat–power technologies.

It is cheaper to reuse the low-grade heat as a source of energy for on-site heating process or power generation (Elson and Hampson, 2015). However, this energy system technique has been hampered by associated problems with storage and transmission distance. Hence, waste heat utilization is suitable and opted for off-site use (Zhang et al., 2016). Due to the specific benefit of varied cycles that can adjust to diverse medium-low heat sources to produce additional electric energy, a waste heat-to-power (WHP) system is accessible and cost-effective, particularly when the heat source temperature is over 150°C (Kurle et al., 2016). Stepping up from low to high heat quality, depending on the temperature difference to be achieved and the relative

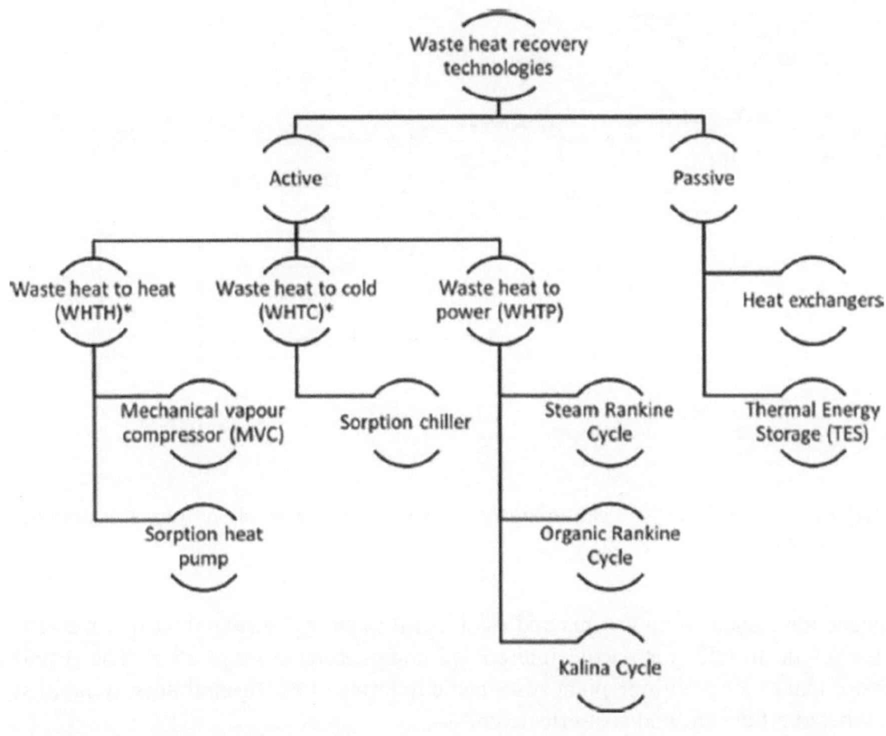

FIGURE 13.1 Classification of waste heat recovery technologies (Brückner et al., 2015).

costs of fuel and electricity, may be more cost-effective. A heat load's temperature is slightly higher than waste heat's low temperature. A heat pump, rather than burning extra fossil fuels, can occasionally generate more efficient waste heat (Incorporated, 2008).

13.2.1.1 Waste Heat-to-Power (WHP) Technologies

The WHP technology is a principal area of applications of WHR. Among the most recent and pronounced WHP techniques for industrial low-grade waste heat recovery applications are thermoelectric generators (TEGs) and thermal power cycles (combined heat and power (CHP)/polygeneration cycle, gas turbine regenerator, and organic Rankine cycle (ORC)).

13.2.1.1.1 Gas Turbine Regenerator

Gas turbine regenerators employ exhaust gases at high temperatures to preheat the compressed air going into the combustion chamber. In gas turbine engines that use low-grade heat that would otherwise be vented into the environment but is used in bottoming cycles, a regenerator is a necessary component. As indicated in Figure 13.2, a heat recovery steam generator (HRSG) is added after the regenerator. With no additional fuel input, the HRSG will trap low-grade heat from the exhaust gas and utilize it in a typical steam Rankine cycle to generate up to 50% more power in the combined cycle plant. Ramakrishnan and Edwards (2016) investigated different

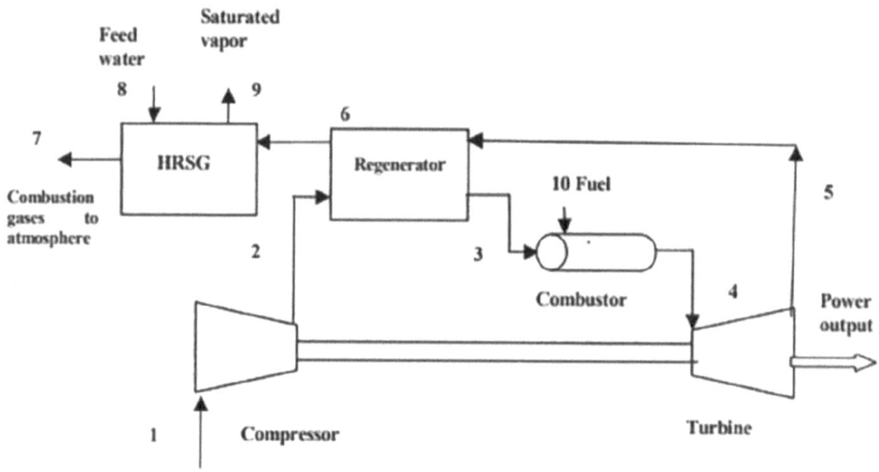

FIGURE 13.2 Schematic of gas turbine cogeneration with regeneration (Ramakrishnan and Edwards, 2016).

regenerative cycle architectures and established them to improve their thermal efficiencies up to 68% for partial intercooled compression configuration. The regenerator makes a gas turbine plant sustainable in terms of eco-friendliness, reliability, improved efficiency, and cost-effectiveness.

13.2.1.1.2 Combined Hybrid Heat & Power/Polygeneration Cycle

The combined heat and power (CHP) cycle and polygeneration cycles (PGCs) pave the way for sustainable development by lowering greenhouse gas (GHG) emissions and reducing reliance on finite, non-renewable fossil fuels. CHP cycles are ideal for generating both heat and electricity at the same time. PGCs, on the other hand, have been utilized to simultaneously generate many (more than two) products such as heating, cooling, power, hydrogen fuel, and salinity. The use of waste heat in CHP and PGC can boost system efficiency by up to 68% while also lowering CO_2 emissions (Al-Sulaiman et al., 2010).

The CHP systems are being recognized as the principal substitute for conventional energy conversion systems such as stand-alone gas turbines or steam power plants from economic, technical, and environmental perspective. Through the integration of two or more power systems, research has demonstrated that the performance of CHP may be improved further for increased efficiency, reduced fuel use, and possibly abating environmental pollution owing to CO_2 emissions. For example, studies on the classification of different types of prime mover integration for hybrid CHP systems such as hybrids of steam and gas turbines CHPs, solar hybrid CHPs, micro-gas turbines, hybrid with fuel cell, and multi-engine CHPs (which include biomass-integrated gasification gas turbine CHP (BIGGT), biomass-integrated gasification combined cycle CHP (BIGCC), and gas turbine CHP (GT)) have been carried out (Mahian et al., 2020).

Recently, some researchers have proposed the integration of renewable energy sources with conventional power systems purposely not only to make the systems more environmentally friendly, but also to conserve the limited available fossil fuel. Akkaya et al. (2008) developed a solid oxide fuel cell/gas turbine CHP model in their work. The performance of the CHP system was investigated using exergy analysis. The study revealed improvement in the exergy efficiency of the power cycle.

Khaljani et al. (2015) investigated an integrated gas turbine and ORC CHP designed to meet the electric power and heating demands. An exergoeconomic analysis was carried out on the CHP system, and it was discovered a reduction in the cost rate of the plant. Zahedi et al. (2019) designed and evaluated the thermodynamic performance of a dual CHP plant powered by a solar parabolic trough collector and a biofuel generator. The study showed improvements in the plant's efficiency and reduction in fuel consumption and CO_2 emission.

The use of hybridized CHPs to combine renewable and synthetic fuels is notable since these technologies reduce environmental pollutions and fuel consumption and improve energy conversion system efficiency.

13.2.1.1.3 Organic Rankine Cycle (ORC)

ORC is recognized as a refined, appropriate, and cost-effective technology for reducing GHG emissions and conserving finite fossil fuel reserves by improving the performance of energy systems that use low-power, low-temperature heat sources (Fakeye et al., 2021). Low-grade heat sources for recovery from various industrial operations (production or manufacture) have the ability to propel ORC systems with output power ranging from roughly 10kW to 10 MW (Auld et al., 2013). Wang et al. (2012), on the other hand, opined that the ORC system is uneconomical when the heat source temperature is less than 100°C.

Because of their clarity and availability of parts, ORCs have been acknowledged as the central low-grade waste heat recovery systems (Ziviani et al., 2014). When working with low-grade heat sources that have acceptable working fluids and operational conditions, the superiority of ORC over other systems such as a steam Rankine cycle is clearly demonstrated (Dhar et al., 2017). At lower temperatures, ORCs, according to Abadi et al. (2014), are more favorable than the steam Rankine cycle. This is because, when employing low-boiling organic fluids to recover low-grade heat at temperatures less than 300°C as bottoming cycles, ORCs produce less than 20 MW of power. ORC systems are also preferred when employing organic working fluids since they only require single-stage expanders, which are easier to install and maintain (Chen et al., 2011).

13.2.1.1.4 Thermoelectric Generators (TEGs)

TEGs are devices that can turn waste heat into power. A TEG is made up of thermocouples that are electrically connected in series, but thermally coupled in parallel and are made up of semiconductors in thermocouples. The device works on the Seebeck effect and can be used in a wide temperature range. Even at high temperatures, the device's conversion efficiency is very low. For example, at temperatures above 1000°C, the efficiency is less than 20%, while at 400°C, it is less than 10% (Bendig, 2015).

TEGs are advantageous in WHR because of their simple assembly, lack of moving parts, lack of noise and vibration, solid-state operation without chemical reactions or harmful residuals, and low maintenance procedures. In addition, the device's tiny size, as well as its ability to establish and promote uniform performance standards, has been identified as a key consideration in its use in industries.

13.3 WASTE HEAT RECOVERY'S ECONOMIC AND ENVIRONMENTAL BENEFITS

The twofold benefits of WHR technologies are principally on cost of saving energy and abatement of environmental pollution. From financial point of view, WHR offers such benefits as appreciable development in energy use performance in energy generation and industrial processes. This leads to considerable conservation in the cost of fuel, while from the environmental point of view, there is appreciable reduction in consumption rate of fossil fuel and discharge of toxic pollutants to the environment. Hence, this furthers sustainability. As a result, low-grade heat is a major source of energy that can be recovered using a variety of low-heat recovery technologies. Industrial and power facilities, for example, have a significant impact on emissions and fuel usage. A 0.1% gain in efficiency for large combustion power plants reduces CO_2 emissions by 1000–1500 tons/year, according to Ekama et al. (2003). Table 13.1 lists the specific advantages of implementing specific WHR systems.

13.3.1 Limits to Effective WHR System Deployment in Developing Countries

WHR technologies have substantial potential, yet their effective deployment in developing nations has been hampered. Some hindrances to the effective implementation of WHP systems in developing nations have been identified (Elson and Hampson, 2015). These include technical gremlins, business cogitations, and policy issues, which are typically intertwined and necessitate separate transactions relating to

TABLE 13.1
The Advantages of Implementing Some WHR Technologies

S. No.	Application	Fuel Abatement (%)	Performance Improvement
1	CHP with ORC, Kalina cycle or transcritical CO_2 cycle (Demirbas, 2016)	About 15–35	About 80% electrical efficiency
2	Combined cycle heat recovery generator in regenerative gas turbine power plant (Iqbal et al., 2017)	Not available	More than 65%
3	CO_2 sequestration	At no extra fuel	Better eco-friendliness
6	Salt removal unit for thermal power plants (Miró et al., 2016)	Up to 1844 kW/h on a 47 MW steam turbine	Not applicable
7	TEG-ORC integrated regenerative system (Shu et al., 2012)	Not available	About 41%–45.71%

the cost-benefits or technical performance of WHR alternatives (Kurle et al., 2016). These barriers include the following.

13.3.1.1 Technical Gremlins

Among the technical gremlins to the development of WHR in developing nations are the following:

- Fluctuating and erratic state of the low-temperature heat source.
- The eccentricity of each low-temperature heat source, for example, temperature and contaminants.
- Insufficient pertinent information and figure.
- Insufficient skills and technical knowledge on technologies.
- The high cost of incorporating a WHR unit into an existing energy conversion system.
- Lack of support for technical training from the government.
- Competition between alternatives to renewable energy (RE).

13.3.1.2 Business Barriers

Among the business barriers to the sustainable development of WHR technologies in developing nations are as follows:

- A desire for quick profits and a reluctance to invest in projects that do not increase output or revenue.
- Lack of financial capacity to purchase, operate, and maintain WHR projects.
- Perceived high unpredictability and unreliability of WHR projects.
- Delay in payback time and high expectations of a too high rate of return.
- Difficulty in securing finance for WHP projects.
- Lack of understanding about the potential and benefits of WHR projects.

13.3.1.3 Policy, Legislative, and Regulatory Barriers

The following policy and regulatory barriers contribute to the sustainable development of WHR technologies in developing nations:

- Lack of viable goads for WHR projects.
- Lack of regulatory support for the WHR projects to thrive in developing nations.
- Lack of consideration for WHR technologies as a renewable energy source.
- Slow perceptibility to the grave consequences of discharge of low-temperature heat to the environment from industrial processes.
- Individual countries' failure to internalize key global policy.

The prime possibility for determining the effective adoption of WHR lies in economic considerations relating to equipment cost and the prediction of its energy savings capacity. However, policymakers, government roles, and energy sector stakeholders' regulations and policies significantly impact WHR project economics' success. Insufficient skills and expertise with prospecting, design, and operation of different plants in the sectors with prospective resources have delayed the growth of industrial waste heat use for power generating and process industries in developing countries (Jung et al., 2014).

13.3.1.4 Possible Remedies to Sustainable Development and Effective Utilization of WHR Systems in Developing Countries

The following are some of the possible solutions for attaining sustainable development and effective use of WHR projects in the goal of global sustainable energy development in poor countries:

- Increased long-term research and development, as well as the use of WHR technologies.
- Creating motivation and encouragement measures, as well as loan facilitation, for the deployment of WHR systems.
- Incorporating regulations that will execute worldwide best practices on the efficient use of WHR systems as an energy-saving technique in developing countries' manufacturing and electricity sectors.
- Creating a favorable climate for worldwide public knowledge and trust in WHR systems as long-term solutions.
- Transfer of technology from developed to developing countries.

13.4 WHR TECHNOLOGIES FOR SUSTAINABILITY AND ECONOMIC GROWTH IN DEVELOPING ECONOMIES

Many terminologies have been used to define sustainable energy in the literature. This tends to overemphasize or trade off sustainability's underpinnings. In a literal sense, "sustainability" means "must be sustained for a long time without interruption or weakening and must remain workable," but not necessarily indefinitely. Hence, "sustainability" refers to an unbiased, trilateral connection of energy, environment, and economics (Vidadili et al., 2017).

In reality, the power industry is currently battling with unpredictability about the consequences of the fast depletion rate, with the correlated environmental pollution of inefficient consumption of finite conventional fossil fuels, the technical and financial hurdles of renewable energy sources, and the proposition of producer gas as a favorable viaduct to renewable energy sources to prolong the life expectancy of determinate conventional fossil fuels (Furlan and Mortarino, 2018). Nevertheless, the present progress on renewable energy sources accounts for only 14% of total global energy consumption (Demirbas, 2016). It grows at a rate of 2.6% per year (EIA, 2016), making the shift to such facilities insufficient to fulfill future energy demand forecasts. Furthermore, solid considerations demand that more than half of the world's non-renewable energy supplies, particularly coal, the cheapest and most abundant of all fossil fuels, be left untouched (Warner and Jones, 2017). For developing countries to leave their fossil resources untapped in pursuit of RE is very unlikely, noting that the 14 members of the OPEC accounting for more than 73% of the world's proven oil reserves are generally categorized as developing nations (Sepehrdoust, 2018).

Global temperature rises should be kept well below 2°C this century, preferably 1.5°C, according to the present Paris climate change agreement. However, based on world population projections, limiting global warming to less than 2°C will necessitate a significant drop in fossil fuel consumption by mid-century (Jones and Warner, 2016). Regardless of climate goals, achieving these goals will necessitate a 37-fold increase in RE contribution from the 2014 level of 13×10^{18} J by 2028, and 75–81 times by 2100 based on the quantitative projected peak of non-renewable energy resources, according to United Nations 2015 population projections (Jones and Warner, 2016).

Hence, the energy mix generally shifts towards cleaner, lower carbon fuels, driven by environmental needs and technological advances (Consumption et al., 2017). Primary energy consumption increased by 2.9% in 2018, about twice its 10-year average of 1.5% yearly (BP Statistical Review of World Energy, 2019). The growth rates of all the fuels were faster than their 10-year averages except for renewable energies (RE). Strikingly, three of the world-leading economic countries, namely China, the USA, and India, collectively accounted for almost 2% of the increase in primary energy consumption. In contrast, the rest of the world accounted for only 0.9%. China is the world's largest energy consumer and CO_2 emitter (Tanaka, 2010), and its CO_2 emissions are continually rising rapidly (Zhang et al., 2012). Despite the present worldwide resolve to transition to renewable energy sources, China's economy, which is rising at a rate of 7% yearly, is fueled by less expensive fossil fuels such as coal (Chen, 2018). Between 1980 and 2012, Amri (2017) studied the relationship between economic growth and energy use in Algeria, a developing country with considerable fossil fuel reserves. Only non-renewable energy and capital, according to the research, were able to stimulate significant economic growth, while renewable energy had a small impact.

The current trend of unsustainable resource deployment will swiftly exhaust known reserves in fossil fuel-producing developing nations such as Pakistan and Nigeria, which rely on the resource to provide a large portion of their energy demands (Aized et al., 2016). Cash and McCormack (2017) highlighted some fundamental considerations concerning the path to a clean energy future, such as how swift and fair the transition to full deployment of renewable energy sources should be, and provided a technique that meets the study's goal. With the current rate of renewable energy research and deployment around the world, it is unlikely that more than half of non-renewable energy sources will go untapped. On the one hand, developing countries, like many others, would choose the path of successfully utilizing sustainable technology to extend the lifecycle of resources while still allowing for the discovery of new reserves and unpredictable technological innovation such as fracking (for shale gas extraction). Global proven oil reserves, on the other hand, grew by 0.9% in 2016 and are expected to last 50.6 years at 2016 production levels (BP, 2017).

In a nutshell, the Paris Agreement identifies sustainable technology as a vital approach to rapidly decarbonizing the energy system. This is especially true for the generation of ecologically friendly electrical energy, and it went on to establish a framework to help developing countries and the world's most vulnerable countries achieve their environmental goals. As a result, achieving a stable, long-term, and low-carbon path is crucial for long-term energy development, particularly in the industrial and power sectors. Economic growth as measured by GDP has been "uncoupled" from total primary energy supply (TPES) and CO_2 emissions in all regions of the world in a study between 1990 and 2014 with the introduction of the concepts of energy intensity and carbon intensity given as TPES/GDP and CO2/GDP, respectively. The study by Goldemberg (2019) showed that carbon footprints are growing faster than TPES globally. Still, the report showed many countries that developed without increasing CO_2 emissions, concluding that there is no basis to believe it is impossible to achieve in many developing countries.

WHR technologies are cost-effective, long-lasting solutions that have acquired widespread acceptance in recent years. It's an appealing way to accomplish sustainable energy development goals by enhancing the overall efficiency of thermal and process facilities, lowering fuel costs and consumption, and reducing hazardous emissions to acceptable levels. Globally, low-grade heat supplies are underutilized, with Canada leading the way in waste heat usage. The following countries' industrial waste heat per energy consumed by industry: the UK (4.2%), EU (5.6%), Germany (5.8%), Belgium (7.5%), Romania (8.0%), Lithuania (10.4%), Slovak Republic (10.5%), Turkey (17.4%), and Canada (26.4%) (Miró et al., 2015).

13.5 CONCLUSIONS

In this chapter, many WHR technologies have been studied, rated, and determined to be adequate for attaining sustainable energy development in poor countries through more efficient use of fossil-based fuels. As previously stated, a suitable technology can result in considerable monetary and environmental benefits, such as reduced primary energy consumption, lower CO_2 emissions, and financial rewards.

There are abundant fossil fuels and waste heat resources that may be used to meet the global energy demand without risking sustainable development until the major difficulties impeding full utilization of renewable energy resources in all parts of the world are resolved. The most significant roadblocks are related to business and regulatory challenges rather than technical issues. As a result, policies that promote beneficial business considerations are urgently needed. As a result, more WHR will be exploited, leading to more sustainable development in the long run. Because of the benefits they give, WHR systems require minimal maintenance and no additional fuel input. According to previous study, the payback period has always been less than 7 years. However, in order to anticipate the predicted lifespan of the confirmed stockpiles of constrained fossil fuels as well as the ecosystem's stability, a thorough assessment of the projected optimum energy usage of specific selected nations in every corner of the world is needed. The case for fully utilizing the many essential WHR technologies for long-term development will be strengthened by such a research.

REFERENCES

Abadi, M., Hooshmand, P., & Behrooz Khezri, A. R. (2014). *Investigation of Using Different Fluids for Using in Gas Turbine-Rankine Cycle*, 1(2), 74–81.

Aized, T., Shahid, M., Bhatti, A. A., Saleem, M., & Anandarajah, G. (2016). Energy security and renewable energy policy analysis of Pakistan. *Renewable and Sustainable Energy Reviews*, 84(March 2017), 155–69. https://doi.org/10.1016/j.rser.2017.05.254.

Akkaya, A. V., Sahin, B., & Huseyin Erdem, H. (2008). An analysis of SOFC/GT CHP system based on exergetic performance criteria. *International Journal of Hydrogen Energy*, 33(10), 2566–77.

Al-Sulaiman, F.A., Dincer, I., & Hamdullahpur, F. (2010). Exergy analysis of an integrated solid oxide fuel cell and organic Rankine cycle for cooling, heating and power production. *Journal of Power Sources*, 195(8), 2346–2354.

Amri, F. (2017). The relationship amongst energy consumption (renewable and non-renewable), and GDP in Algeria. *Renewable and Sustainable Energy Reviews*, 76, 62–71. https://doi.org/10.1016/j.rser.2017.03.029.

Auld, A., Berson, A., & Hogg, S. (2013). Organic rankine cycles in waste heat recovery: A comparative study. *International Journal of Low-Carbon Technologies*, 8(suppl 1), 9–18. https://doi.org/10.1093/ijlct/ctt033.

Bao, J., & Zhao, L. (2013). A review of working fluid and expander selections for organic Rankine cycle. *Renewable and Sustainable Energy Reviews*, 24, 325–342. https://doi.org/10.1016/j.rser.2013.03.040.

Bendig, M. (2015). *Integration of Organic Rankine Cycles for Waste Heat Recovery in Industrial Processes*, EPFL, Lausanne. PAR. 6536.

Berthou, M., & Bory, D. (2012). *Overview of Waste Heat in the Industry in France*. 453–459.

Bianchi, G., Mcginty, R., Oliver, D., Brightman, D., Zaher, O., Tassou, S. A., et al. (2017). Development and analysis of a packaged Trilateral Flash Cycle system for low grade heat to power conversion applications. *Thermal Science and Engineering Progress*, 4, 113–121. https://doi.org/10.1016/j.tsep.2017.09.009.

Bombarda, P., Invernizzi, C. M., & Pietra, C. (2010). Heat recovery from Diesel engines: A thermodynamic comparison between Kalina and ORC cycles. *Applied Thermal Engineering*, 30(2–3), 212–219. https://doi.org/10.1016/j.applthermaleng.2009.08.006.

BP Statistical Review of World Energy. (2019).

BP. (2017). BP statistical review of world energy 2017. *British Petroleum*, 66, 1–52. Retrieved from http://www.bp.com/content/dam/bp/en/corporate/pdf/energy-economics/statistical-review-2017/bp-statistical-review-of-world-energy-2017-full-report.pdf.

Brückner, S., Liu, S., Miró, L., Radspieler, M., Cabeza, L. F., & Lävemann, E. (2015). Industrial waste heat recovery technologies: An economic analysis of heat transformation technologies. *Applied Energy*, 151, 157–167. https://doi.org/10.1016/j.apenergy.2015.01.147.

Brueckner, S., Miró, L., Cabeza, L. F., Pehnt, M., & Laevemann, E. (2014). Methods to estimate the industrial waste heat potential of regions – A categorization and literature review. *Renewable and Sustainable Energy Reviews*, 38, 164–171. https://doi.org/10.1016/j.rser.2014.04.078.

Cash, D. W., & McCormack, J. W. (2017). Choices on the road to the clean energy future. *Energy Research and Social Science*, 35, 224–226. https://doi.org/10.1016/j.erss.2017.10.035.

Chen, H., Yogi Goswami, D., Rahman, M. M., & Stefanakos, E. K. (2011). Energetic and exergetic analysis of CO_2- and R32-based transcritical Rankine cycles for low-grade heat conversion. *Applied Energy*, 88(8), 2802–2808. https://doi.org/10.1016/j.apenergy.2011.01.029.

Chen, S., Guo, Z., Liu, P., & Li, Z. (2018). Advances in clean and low-carbon power generation planning. *Computers and Chemical Engineering*, 116, 296–305. https://doi.org/10.1016/j.compchemeng.2018.02.012.

Chen, Y. (2018). Factors influencing renewable energy consumption in China: An empirical analysis based on provincial panel data. *Journal of Cleaner Production*, 174, 605–615. https://doi.org/10.1016/j.jclepro.2017.11.011.

Cipollone, R., Bianchi, G., Di Bartolomeo, M., Di Battista, D., & Fatigati, F. (2017). Low grade thermal recovery based on trilateral flash cycles using recent pure fluids and mixtures. *Energy Procedia*, 123, 289–296. https://doi.org/10.1016/j.egypro.2017.07.246.

Date, A., Alam, F., Khaghani, A., & Akbarzadeh, A. (2012). Investigate the potential of using trilateral flash cycle for combined desalination and power generation integrated with salinity gradient solar ponds. *Procedia Engineering*, 49, 42–49. https://doi.org/10.1016/j.proeng.2012.10.110.

Demirbas, A. (2016). *Waste Energy for Life Cycle Assessment*. https://doi.org/10.1007/978-3-319-40551-3.

Dhar, H., Kumar, S., & Kumar, R. (2017). A review on organic waste to energy systems in India. *Bioresource Technology*, 245(Part A), 1229–1237. https://doi.org/10.1016/j.biortech.2017.08.159.

EIA. (2016). World energy demand and economic outlook EIA's handling of non-U.S. policies in the International Energy Outlook. *U.S. Energy Information Administration*, 2016(May 2015), 7–17. Retrieved from http://www.eia.gov/forecasts/ieo/world.cfm.

Ekama, G. A. (2003). Using bioprocess stoichiometry to build a plant-wide mass balance based steady-state WWTP model. *Water Research*, 43(8), 2101–2120. https://doi.org/10.1016/j.watres.2009.01.036.

Elson, A., & Hampson, A. (2015). *Waste Heat to Power Market Assessment*.

Fakeye, A. B. & Oyedepo, S. O. (2019), Designing optimized organic rankine cycles systems for waste heat-to-power conversion of gas turbine flue gases. *Journal of Physics: Conference Series*, 1378(2019), 032097. https://doi.org/10.1088/1742-6596/1378/3/032097.

Fakeye, A. B., Oyedepo, S. O., Fayomi, O. S. I., Dirisu, J. O., & Udoye, N. E. (2021). Fossil fuel combustion, conversion to near-zero waste through organic rankine cycle. In C. M. Hussain, P. Di Sia (eds.), *Handbook of Smart Materials, Technologies, and Devices*. https://doi.org/10.1007/978-3-030-58675-1_69-1. pp. 1–19.

Feher, E. G. (1968). The supercritical thermodynamic power cycle. *Energy Conversion*, 8(2), 85–90.

Feng, Y., Hung, T. C., Zhang, Y., Li, B., Yang, J., & Shi, Y. (2015). Performance comparison of low-grade ORCs (organic Rankine cycles) using R245fa, pentane and their mixtures based on the thermoeconomic multi-objective optimization and decision makings. *Energy*, 93(2015), 2018–2029. https://doi.org/10.1016/j.energy.2015.10.065.

Fiaschi, D., Manfrida, G., Rogai, E., & Talluri, L. (2017). Exergoeconomic analysis and comparison between ORC and Kalina cycles to exploit low and medium-high temperature heat from two different geothermal sites. *Energy Conversion and Management*, 154(August), 503–516. https://doi.org/10.1016/j.enconman.2017.11.034.

Fischer, J. (2011). Comparison of trilateral cycles and organic Rankine cycles. *Energy*, 36(10), 6208–6219. https://doi.org/10.1016/j.energy.2011.07.041.

Franjo, D., Romagnoli, A., Fox, T., & Schrøder, A. (2017). Potential of waste heat and waste cold energy recovery in Singapore for district cooling applications: Impacts on energy system. *Proceedings of the 40th IAEE International Conference – Extended abstracts: Meeting the Energy Demands of Emerging Economies*. IAEE.

Furlan, C., & Mortarino, C. (2018). Forecasting the impact of renewable energies in competition with non-renewable sources. *Renewable and Sustainable Energy Reviews*, 81(June 2017), 1879–1886. https://doi.org/10.1016/j.rser.2017.05.284.

Huang, S., & Li, C. (2017). *An Improved System for Utilizing Low-Temperature Waste Heat of Flue Gas from Coal-Fired Power Plants*. https://doi.org/10.3390/e19080423.

Iglesias, S., Ferreiro, R., Carbia, J., & Iglesias, D. (2017). Critical review of the first-law efficiency in different power combined cycle architectures. *Energy Conversion and Management*, 148, 844–859. https://doi.org/10.1016/j.enconman.2017.06.037.

Incorporated, B. (2008). Waste heat recovery: Technology opportunities in the US industry. *Waste Heat Recovery: Technology Opportunities in the US Industry*, 1–112. https://doi.org/10.1017/CBO9781107415324.004.

Iqbal, M. A., Ahmadi, M., Melhem, F., Rana, S., Akbarzadeh, A., & Date, A. (2017). Power generation from low grade heat using trilateral flash cycle. *Energy Procedia*, 110(December 2016), 492–497. https://doi.org/10.1016/j.egypro.2017.03.174.

Jones, G. A., & Warner, K. J. (2016). The 21st century population-energy-climate nexus. *Energy Policy*, 93, 206–212. https://doi.org/10.1016/j.enpol.2016.02.044.

Jung, H. C., Krumdieck, S., & Vranjes, T. (2014). Feasibility assessment of refinery waste heat-to-power conversion using an organic Rankine cycle. *Energy Conversion and Management*, 77, 396–407. https://doi.org/10.1016/j.enconman.2013.09.057.

Kandathil, A. K. (2016). *A Guide to Working Fluid Selection for Organic Rankine Cycle ORC Generators. Genixx, Heatcatcher.* Retrieved from http://www.heatcatcher.com/guide-working-fluid-selection-organic-rankine-cycle-orc-generators/.

Khaljani, M., Khoshbakhti, S. R. & Bahlouli, K. (2015). Comprehensive analysis of energy, exergy and exergo-economic of cogeneration of heat and power in a combined gas turbine and organic Rankine cycle. *Energy Conversion and Management*, 97, 154–65.

Kurle, D., Schulze, C., Herrmann, C., & Thiede, S. (2016). Unlocking waste heat potentials in manufacturing. *Procedia CIRP*, 48, 289–294. https://doi.org/10.1016/j.procir.2016.03.107.

Lai, N. A., & Fischer, J. (2012). Efficiencies of power flash cycles. *Energy*, 44(1), 1017–1027. https://doi.org/10.1016/j.energy.2012.04.046.

Landelle, A., Tauveron, N., Revellin, R., Haberschill, P., & Colasson, S. (2017). Experimental investigation of a transcritical organic rankine cycle with scroll expander for low – Temperature waste heat recovery. *Energy Procedia*, 129, 810–817. https://doi.org/10.1016/j.egypro.2017.09.142.

Lee, C. E., Yu, B., & Lee, S. (2015). An analysis of the thermodynamic efficiency for exhaust gas recirculation-condensed water recirculation-waste heat recovery condensing boilers (EGR-CWR-WHR CB). *Energy*, 86, 267–275. https://doi.org/10.1016/j.energy.2015.04.042.

Li, M., Wang, J., Li, S., Wang, X., He, W., & Dai, Y. (2014). Thermo-economic analysis and comparison of a CO2 transcritical power cycle and an organic Rankine cycle. *Geothermics*, 50, 101–111. https://doi.org/10.1016/j.geothermics.2013.09.005.

Li, Y. (2013). *Analysis of Low Temperature Organic Rankine Cycles for Solar Applications Analysis of Low Temperature Organic Rankine Cycles for Solar Applications.* Theses and Dissertations, 1113.

Mahian, O., Mirzaie, M.R., Kasaeian, A., and Mousavi, S.H. (2020), Exergy analysis in combined heat and power systems: A review. *Energy Conversion and Management*, 226, 113467.

Mckenna, R. C., & Norman, J. B. (2010). Spatial modelling of industrial heat loads and recovery potentials in the UK. *Energy Policy*, 86, 267–275. https://doi.org/10.1016/j.enpol.2010.05.042.

Miró, L., Brückner, S., & Cabeza, L. F. (2015). Mapping and discussing Industrial Waste Heat (IWH) potentials for different countries. *Renewable and Sustainable Energy Reviews*, 51, 847–855. https://doi.org/10.1016/j.rser.2015.06.035.

Miró, L., Brueckner, S., Mckenna, R., & Cabeza, L. F. (2016). Methodologies to estimate industrial waste heat potential by transferring key figures: A case study for Spain. *Applied Energy*, 169, 866–873. https://doi.org/10.1016/j.apenergy.2016.02.089.

Miró, L., Gasia, J., & Cabeza, L. F. (2016). Thermal energy storage (TES) for industrial waste heat (IWH) recovery: A review. *Applied Energy*, 179, 284–301. https://doi.org/10.1016/j.apenergy.2016.06.147.

Oyedepo, S.O., & Fakeye, A.B. (2020). Electric power conversion of exhaust waste heat recovery from gas turbine power plant using organic Rankine cycle. *International Journal of Energy and Water Resources*, 4, 139–150. https://doi.org/10.1007/s42108-019-00055-3.

Panayiotou, P., Bianchi, G., Georgiou, G., Aresti, L., Argyrou, M., Agathokleous, R., et al. (2017). Preliminary assessment of waste potential District heat in major European industries. *Energy Procedia*, 86, 267–275. https://doi.org/10.1016/j.egypro.2017.07.263.

Peris, B., Navarro-Esbrí, J., Molés, F., & Mota-Babiloni, A. (2015). Experimental study of an ORC (organic Rankine cycle) for low grade waste heat recovery in a ceramic industry. *Energy*, 85, 534–542. https://doi.org/10.1016/j.energy.2015.03.065.

Rahbar, K., Mahmoud, S., Al-Dadah, R. K., Moazami, N., & Ashmore, D. (2017). Feasibility study of power generation through waste heat recovery of wood-burning stove using the ORC technology. *Sustainable Cities and Society*. https://doi.org/10.1016/j.scs.2017.09.013.

Ramakrishnan, S., & Edwards, C. F. (2016). Maximum-efficiency architectures for heat- and work-regenerative gas turbine engines. *Energy*, 100, 115–128. https://doi.org/10.1016/j.energy.2016.01.044.

Reis, M. M. L., & Gallo, W. L. R. (2018). Study of waste heat recovery potential and optimization of the power production by an organic Rankine cycle in an FPSO unit. *Energy Conversion and Management*, 157(September 2017), 409–422. https://doi.org/10.1016/j.enconman.2017.12.015.

Rocco, M. V., Ferrer, R. J. F., & Colombo, E. (2018). Understanding the energy metabolism of world economies through the joint use of production- and consumption-based energy accountings. *Applied Energy*, 211(November 2017), 590–603. https://doi.org/10.1016/j.apenergy.2017.10.090.

Sepehrdoust, H. (2018). Kasetsart Journal of Social Sciences Impact of information and communication technology and fi nancial development on economic growth of OPEC developing economies. *Kasetsart Journal of Social Sciences*, 6–11. https://doi.org/10.1016/j.kjss.2018.01.008.

Shu, G., Zhao, J., Tian, H., Liang, X., & Wei, H. (2012). Parametric and exergetic analysis of waste heat recovery system based on thermoelectric generator and organic Rankine cycle utilizing R123. *Energy*, 45(1), 806–816. https://doi.org/10.1016/j.energy.2012.07.010.

Singh, D. V., & Pedersen, E. (2016). A review of waste heat recovery technologies for maritime applications. *Energy Conversion and Management*, 111(X), 315–328. https://doi.org/10.1016/j.enconman.2015.12.073.

Tan, Y., Li, X., Zhao, L., Li, H., Yan, J., & Yu, Z. (2014). Study on utilization of waste heat in cement plant. *Energy Procedia*, 61, 455–458. https://doi.org/10.1016/j.egypro.2014.11.1147.

Thermax. (2015). *Waste Heat Recovery for Chocolate Cooling at a Leading Confectionery Company*.

Thornley, P., & Walsh, C. (2010). *Addressing the Barriers to Utilisation of Low Grade Heat from the Thermal Process Industries*.

Upathumchard, U. (2014). *Waste Heat Recovery Options in a Large Gas- Turbine Combined Power Plant Waste Heat Recovery Options in a Large Gas-Turbine Combined Power Plant*.

Varga, Z., & Palotai, B. (2017). Comparison of low temperature waste heat recovery methods. *Energy*. https://doi.org/10.1016/j.energy.2017.07.003.

Vazhappilly, C. V., Tharayil, T. & Nagarajan, A. P. (2013). Modeling and experimental analysis of generator in vapour absorption refrigeration system. *International Journal of Engineering Research and Applications*, 3(5), 63–67.

Vidadili, N., Suleymanov, E., Bulut, C., & Mahmudlu, C. (2017). Transition to renewable energy and sustainable energy development in Azerbaijan. *Renewable and Sustainable Energy Reviews*, 80(June), 1153–1161. https://doi.org/10.1016/j.rser.2017.05.168.

Wang, Y., Tang, Q., Wang, M., & Feng, X. (2017). Thermodynamic performance comparison between ORC and Kalina cycles for multi-stream waste heat recovery. *Energy Conversion and Management*, 143, 482–492. https://doi.org/10.1016/j.enconman.2017.04.026.

Wang, Z. Q., Zhou, N. J., Guo, J., & Wang, X. Y. (2012). Fluid selection and parametric optimization of organic Rankine cycle using low temperature waste heat. *Energy*, 40(1), 107–115. https://doi.org/10.1016/j.energy.2012.02.022.

Warner, K. J., & Jones, G. A. (2017). A population-induced renewable energy timeline in nine world regions. *Energy Policy*, 101(August 2016), 65–76. https://doi.org/10.1016/j.enpol.2016.11.031.

Yue, C., Han, D., Pu, W., & He, W. (2015). Comparative analysis of a bottoming transcritical ORC and a Kalina cycle for engine exhaust heat recovery. *Energy Conversion and Management*, 89, 764–774. https://doi.org/10.1016/j.enconman.2014.10.029.

Zahedi, A., Timasi, H., Kasaeian, A., & Mirnezami, S. A. (2019). Design and construction of a new dual CHP-type renewable energy power plant based on an improved parabolic trough solar collector and a biofuel generator. *Renewable Energy*, 135, 485–95.

Zeb, K., Ali, S. M., Khan, B., Mehmood, C. A., Tareen, N., Din, W., et al. (2017). A survey on waste heat recovery: Electric power generation and potential prospects within Pakistan. *Renewable and Sustainable Energy Reviews*, 75(July 2016), 1142–1155. https://doi.org/10.1016/j.rser.2016.11.096.

Zhang, D., Ma, L., Liu, P., Zhang, L., & Li, Z. (2012). A multi-period superstructure optimisation model for the optimal planning of China's power sector considering carbon dioxide mitigation. Discussion on China's carbon mitigation policy based on the model. *Energy Policy*, 41, 173–183. https://doi.org/10.1016/j.enpol.2011.10.031.

Zhang, X., Wu, L., Wang, X., & Ju, G. (2016). Comparative study of waste heat steam SRC, ORC and S-ORC power generation systems in medium-low temperature. *Applied Thermal Engineering*, 106, 1427–1439. https://doi.org/10.1016/j.applthermaleng.2016.06.108.

Ziviani, D., Beyene, A., & Venturini, M. (2014). Advances and challenges in ORC systems modeling for low grade thermal energy recovery. *Applied Energy*, 121, 79–95. https://doi.org/10.1016/j.apenergy.2014.01.074.

Zoltan, V. & Palotai, B. (2017). Comparison of low temperature waste heat recovery methods. *Energy*, 137, 1286–1292. https://doi.org/10.1016/j.energy.2017.07.003.

Index